The Waxy Secrets of Kermes Oak: Unveiling the Role of Cuticular Wax in Water Management

Ember

General introduction 01

Material and Methods 04

    Sampling sites, plant material, and grown conditions ....

    Morphological and anatomical leaf traits..

    Scanning electron microscopy..

    Minimum leaf water conductance ....

    Chemical composition of cuticular waxes ....

    Chemical composition of the cutin matrix ....

    Leaf photosynthetic thermal tolerance ....

    Statistical analysis ....

Chapter 1 15

Cuticular leaf wax deposition and cuticular transpiration barrier in *Quercus coccifera* L.: does the environment matter?...

1.1 Introduction..

1.2 Results ....

    1.2.1 Leaf traits ....

    1.2.2 Leaf surface properties...

    1.2.3 Leaf minimum conductance..

    1.2.4 Chemical composition of leaf cuticular waxes..

1.3 Discussion..

Chapter 2 29

Comparative analysis of temperature effects on the cuticular transpiration barrier of the desert water-spender *Citrullus colocynthis* and the water-saver *Phoenix dactylifera**..........

2.1 Introduction..

2.2 Results ....

    2.2.1 Morphological and anatomical leaf traits...

    2.2.2 Leaf surface properties...

    2.2.3 Leaf drying curves and minimum leaf water conductance...

    2.2.4 Chemical composition of cuticular waxes ....

    2.2.5 Chemical composition of the cutin matrix..

2.3 Discussion ....

Chapter 3 .... **53**
Coevolution of photosynthetic thermal tolerance and cuticular transpiration barrier in non-succulent xerophytes ....

3.1 Introduction ....
3.2 Results ....
    3.2.1 Minimum leaf conductance ....
    3.2.2 Leaf photosynthetic thermal tolerance ....
    3.2.3 Leaf photosynthetic thermal tolerance and cuticular transpiration barrier ....
    3.2.4 Chemical analysis of the cuticular leaf waxes ....
    3.2.5 Cuticular leaf wax of *Byrsonima gardneriana* (tropical dry forest) ....
    3.2.6 Cuticular leaf wax of *Calotropis procera* (hot desert) ....
    3.2.7 Cuticular leaf wax of *Ceratonia siliqua* (Mediterranean woodland) ....
    3.2.8 Cuticular leaf wax of *Chamaerops humilis* (Mediterranean woodland) ....
    3.2.9 Cuticular leaf wax of *Citrullus colocynthis* (hot desert) ....
    3.2.10 Cuticular leaf wax of *Cynophalla flexuosa* (tropical dry forest) ....
    3.2.11 Cuticular leaf wax of *Olea europaea* var. sylvestris (Mediterranean woodland) ..
    3.2.12 Cuticular leaf wax of *Phillyrea angustifolia* (Mediterranean woodland) ....
    3.2.13 Cuticular leaf wax of *Phoenix dactylifera* (hot desert) ....
    3.2.14 Cuticular leaf wax of *Pistacia lentiscus* (Mediterranean woodland) ....
    3.2.15 Cuticular leaf wax of *Quercus coccifera* (Mediterranean woodland) ....
    3.2.16 Cuticular leaf wax of *Senna italica* (hot desert) ....
    3.2.17 Cuticular leaf wax of *Smilax aspera* (Mediterranean woodland) ....
    3.2.18 Cuticular leaf wax of *Smilax japicanga* (tropical dry forest) ....
3.3 Discussion ....

General discussion ....
    The cuticular permeability of non-succulent xerophytes ....
    The plant cuticle as leaf cuticular transpiration barrier in hot arid biomes ....
    The effect of temperature on the cuticular transpiration barrier of non-succulent xerophytes ....

References **108**

General introduction

The first plants colonised the land circa 440 million years ago (Niklas, 1997 and 2016; Lewis and McCourt, 2004). The transition from an exclusively aquatic environment to a terrestrial one exposed the first land plants to the highly desiccating effects of the surrounding atmosphere and sudden temperature fluctuations. The first multicellular plants acquired through evolution a multitude of metabolic, physiological and morphological modifications, which enabled them to cope with the challenges that came up with the new environment (Niklas *et al.*, 2017). One of the most critical adaptive traits allowing plant life in the land is the capability to synthesise, deposit and maintain a hydrophobic surface layer covering the aerial organs – the plant cuticle (Yeats and Rose, 2013).

Plant cuticle plays essential autecological functions. Although the primary function of the plant cuticle is avoiding non-stomatal water loss to the atmosphere, this structure also protects the plants against multiple abiotic (such as excessive UV radiation and wind abrasion) and biotic stresses (for instance pathogen infection and herbivory). Most of the plant water loss occurs through transpiration from the leaves and primary stems. Thus, stomata and cuticle work together to keep up an appropriated plant water status. Water moves from the leaf interior to the atmosphere through the cuticular and stomatal pathways. The cuticular path is fixed, whereas the stomatal one varies according to the opening and closing of the stomata. Drought-stressed plants close the stomata, thereby minimising water loss through the stomatal pathway. However, under such conditions, there is still the inevitable water loss from the mesophyll to the dry atmosphere across the cuticle – the cuticular transpiration. Within arid environments, the combination of low air humidity and high temperatures exacerbates the driving force to water loss. Therefore, the extent of cuticular transpiration in relation to plant water reservoirs and water acquisition has a crucial impact on plant fitness and survival (Schuster *et al.*, 2017), particularly on non-succulent xerophytes.

The statement that xerophilous plants possess a particularly tough cuticle to retard water loss is common found in the literature (Purves *et al.*, 2004; De Mico and Arone, 2012).

studies attempted to categorise cuticular permeances according to the original plant habitat and reported that tropical evergreen epiphytic or climbing plants possessed the lowest permeances, followed by Mediterranean evergreen, and the highest permeance values were found for temperate deciduous plants (Schreiber and Riederer, 1996; Riederer and Schreiber, 2001). Also, at a congeneric level, the cuticular conductances of 11 *Quercus* species growing in a common garden were investigated, and the g_{min} of Mediterranean evergreen oaks were slightly lower than that of temperate deciduous oaks (Gil-Pelegrín *et al.,* 2017). Therefore, one might expect that plants from arid environments have lower permeabilities in comparison with plants from humid regions. Although this seems to be reasonable, few studies have been done to evaluate whether the environmental characteristics can indeed alter the plant cuticular permeability. In chapter one, we investigate whether contrasting environments (the Mediterranean and Temperate climate-types) are capable of modifying the cuticular features of genetically related individuals of *Quercus coccifera* L.

Arid environments correspond to almost one-third of the total land over the world. Plant evolution led to multiple modifications, thereby enabling plants to cope with water shortage and intense solar radiation. The so-called water-saver and water-spender life strategies represent successful evolutionary adaptations allowing plants to deal with the constraints of hot and dry habitats (Smith *et al.,* 1997). Drought avoiding plants present two contrasting strategies in response to a low water source (Baquedano and Castillo, 2006): (1) water-saver plants possess highly efficient stomatal control and prevent damage by closing stomata before the leaf water potential changes substantially; (2) water-spender plants open the stomata, thereby decreasing the leaf water potential allowing them to extract water from the soil to compensate the high water loss by stomatal transpiration (Guehl and Aussenac, 1987; Lo Gullo and Salleo, 1988). Under maximally stomatal closure, the temperature effect on the cuticular water permeability is one of the crucial parameters influencing the physiological and ecological role of the cuticle. Taking account the fast stomatal response of water-saver plants under the high evaporative demand and elevated temperatures; it might be reasonable that these plants also have a more efficient cuticular transpiration barrier. In chapter two, we compare the of the cuticular transpiration barrier of the desert water-saver *Phoenix dactylifera* L. and the water-spender *Citrullus colocynthis* (L.) Schrad. within the range of ecologically relevant temperatures.

Flowering plants are found within the entire continuum of habitats, from those with few weeks of frost-free weather to hot deserts. Plant evolution to occupy dry regions with concomitantly high temperatures required many morphological and physiological features, which include photosynthetic thermal tolerance (Knight and Ackerly, 2003) and traits related to water balance. Photosynthesis promptly responds to temperature (Knight and Ackerly, 2003), and the photosystem II (PSII) is one of the most thermally labile constituents of photosynthesis (Weis and Berry, 1988; Havaux, 1993). Thus, the excitation capacity of PSII represents a non-invasive proxy for accessing the leaf heat tolerance. Similarly, the minimum leaf conductance equals the minimum und inevitable water loss under maximally stomal closure and, therefore, can be used as an efficiency indicator of the cuticular transpiration barrier. In chapter three, we investigate how adaptations to cope with water (at cuticular level) and heat (at photosynthetic level) stresses are related to 14 plant species natives from three hot arid biomes.

Material and Methods

Sampling sites, plant material, and grown conditions

In chapter one, to reduce the within-species genetic variability, we collected *Quercus coccifera* L. seeds from the same open-pollinated mother tree (Himrane *et al.*, 2004) in a natural population of Sistema Ibérico Meridional provenance of Spain. Seeds were sown in 2009 and cultivated following the methodology described in Peguero-Pina *et al.* (2016). Firstly, plants were grown in 0.5 L containers inside a greenhouse with a mixture of 80% compost (Neuhaus Humin Substrate N6; Klasman-Deilmann GmbH, Geeste, Germany) and 20% perlite. Secondly, after the first vegetative period, seedlings were transplanted to 25 L containers with the same mixture proportion of compost:perlite (80:20). After that, plants were cultivated outdoors at CITA de Aragón (41°39'N, 0°52'W, 200m a.s.l. Zaragoza, Spain) under the Mediterranean (MED) climatic conditions (mean annual temperature 15.4°C, total annual precipitation 298 mm) and irrigated every 3-4 days. Finally, ten of these 2-year-old plants were randomly selected and moved to the Jardín Botánico de Iturrarán (43°13'N, 02°01'W, 70 m a.s.l., Gipuzkoa, Spain), which features temperate (TEM) conditions (mean annual temperature 14.5°C, total annual precipitation 1631 mm). A more detailed description of the site climate conditions can be found in Peguero-Pina *et al.*, (2016). Measurements were conducted in 2017 using one-year fully developed leaves of 8-year-old plants living under two different conditions: MED and TEM.

In chapter two, seeds of *Citrullus colocynthis* (L.) Schrad. were harvested in a natural population growing near Riyadh, Saudi Arabia. Circa 5-years-old and 2 m high plants of *Phoenix dactylifera* L. were purchased from a commercial supplier ('Der Palmenmann', Bottrop, Germany). Plants were cultivated in a greenhouse at 24°C/18°C and 50% ± 5% relative humidity day/night under a 14 h light regime at 250 µmol m^{-2} sec^{-1} white light (Philips Master Agro, 400 W). Healthy fully expanded current year leaves of *C. colocynthis* and leaflets (pinnae) from 1-year-old leaves of *P. dactylifera* were harvested and used for the experiments.

In chapter three, plants from three hot arid biomes distributed within three continents across the globe were chosen. The sampling sites comprise Mediterranean shrubland (southern Portugal), a tropical dry forest (northeastern Brazil), and a hot desert (northeastern Saudi Arabia). According to the climate world map of Köppen-Geiger (Peel *et al.*, 2007), the site climates are classified as Csa type (hot-summer Mediterranean) in the Mediterranean shrubland, BSh climate type (hot semi-arid) in the tropical dry forest, and BWh climate type (hot desert) in the desert site (Fig. 1). The 14 flowering species selected for testing our hypothesis comprise seven Mediterranean species, three dry forest species, and four desert species (Table 1). Plant species were chosen according to the availability of vegetative material at the sampling time. Fully expanded leaves were harvested in natural populations growing in the field (except for desert plants) and packed in plastic bags for following analyses.

Seeds of the desert plants *Calotropis procera* (Aiton) Dryand. and *Senna italica* Mill. were obtained from natural populations growing near Riyadh, Saudi Arabia. Seeds were germinated in a greenhouse. The seeds, including those of *C. colocynthis*, were sown in 2015 and germinated in 0.5 L pots with a mixture of 66% compost (Einheitserde Typ P/ED73, Gebrüder Patzer GmbH & Co. KG), 28% autoclaved sand, and 6% perlite inside a greenhouse of the University of Würzburg, Germany. After the first growth cycle, the seedlings were transplanted to 13 L pots filled with the same mixture of compost, sand, and perlite. Fertilizer (Hakaphos® Blau, COMPO Expert GmbH) was supplied when the plants evinced insufficiency of nutrients. Plants were watered always that necessary to maintain a favorable water status. During the winter, plants were maintained in a greenhouse at 24°C/18°C and 50% ± 5% relative humidity day/night under a 14 h light regime at 250 µmol m^{-2} sec^{-1} white light (Philips Master Agro, 400 W) and moved to the outside garden in the summer.

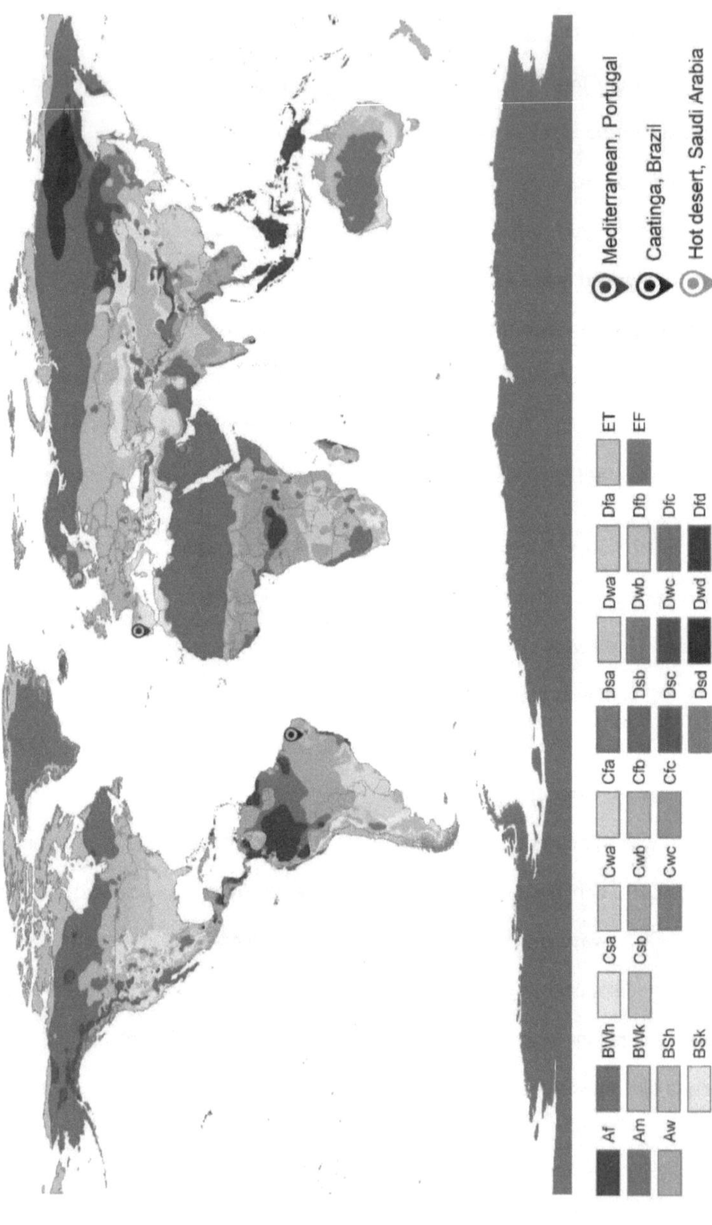

Fig. 1. Sampling sites were drawn in the world map of the Köppen-Geiger climate classification updated by Peel *et al.* (2007). The site climates are classified as hot-summer Mediterranean type (Csa) in the Mediterranean site, hot semi-arid type (BSh) in the Caatinga site, and hot desert type (BWh) in the desert site. See Peel *et al.* (2007) for further legend details.

Table 1. Plant species sampled from three hot and dry biomes.

Plant species	Family	Original habitat	Biome	Location	Latitude	Longitude
Phoenix dactylifera L.	*Arecaceae*					
Calotropis procera (Aiton) Dryand.	*Apocynaceae*	Desert	Hot desert	Asia	25° 24' 35" N	47° 14' 32" E
Senna italica Mill.	*Leguminosae*					
Citrullus colocynthis (L.) Schrad.	*Cucurbitaceae*					
Smilax aspera L.	*Smilacaceae*					
Chamaerops humilis L.	*Arecaceae*					
Ceratonia siliqua L.	*Leguminosae*					
Pistacia lentiscus L.	*Anacardiaceae*	Woodland	Mediterranean	Europe	37° 01' 35" N	08° 54' 35" W
Olea europaea var. sylvestris (Mill.) Lehr	*Oleaceae*					
Quercus coccifera L.	*Fagaceae*					
Phillyrea angustifolia L.	*Oleaceae*					
Smilax japicanga Griseb.	*Smilacaceae*					
Byrsonima gardneriana A.Juss.	*Malpighiaceae*	Tropical dry forest	Caatinga	South-America	08° 34' 52" S	37° 15' 40" W
Cynophalla flexuosa (L.) J.Presl	*Capparaceae*					

7

Morphological and anatomical leaf traits

Before measurement, leaves were rehydrated overnight in a humid chamber. The saturation leaf weights were determined using an analytical balance (MC-1 AC210S, Sartorius; precision 0.1 mg). Dry weight was obtained after oven drying the leaves at 90°C for 24 h. The actual fresh weights during leaf drying experiments were used to calculate the relative water deficit (RWD) according to:

$$RWD = 1 - \frac{actual\ fresh\ weight - dry\ weight}{saturated\ weight - dry\ weight} \tag{1}$$

For leaf area determination, leaves were scanned at high resolution (600 dpi) using a flatbed scanner. The projected leaf area was measured from the scanned leaf image using the Adobe Photoshop image analysis software. Leaf mass per area (LMA) was obtained by dividing the dry weight with the leaf area. The water content per leaf area (WC LA^{-1}) was calculated by dividing the leaf water content with the leaf area.

Scanning electron microscopy

The structure of the leaf cuticular surface was characterised using scanning electron microscopy. Small air-dried pieces (2-3 mm) of fully developed leaves were mounted on aluminium stubs using double-sided adhesive tape and sputter-coated with 10-15 nm gold-palladium (150 s, 25 mA, partial argon pressure 0.05 mbar, SCD005 sputter coater, Bal-Tec). The samples were investigated with a field-emission scanning electron microscope (JEOL JSM-7500F) using a 5-kV acceleration voltage and a 10 mm working distance. Pictures were taken from both adaxial and abaxial leaf surfaces.

Minimum leaf water conductance

Minimum leaf water conductance (g_{min}) of whole leaves represents the "*lowest conductance a leaf can reach when stomata are completely closed as a result of desiccation stress*" (Körner, 1995). The g_{min} was determined gravimetrically from the consecutive weight loss of desiccating leaves in dark and low air humidity (Burghardt and Riederer, 2003). High melting (68°C) paraffin wax (Fluka) was used to seal the wounds of cut petioles of water saturated leaves and leaflets. Subsequently, the sealed leaves/leaflets were placed in an incubator (IPP 110, Memmert) with controlled temperature. Leaf minimum conductance was measured at 25°C, 30°C, 35°C, 40°C, 45°C, and 50°C for evaluating the effect of temperature on the minimum leaf conductance. The air temperature and humidity in the incubator were monitored with a digital thermo-hygrometer (Testoterm 6010, Testo). Silica gel (Applichem) was used to control the air humidity in the incubator. The actual weight of desiccating leaves was determined as a function of desiccation time using a high precision balance (MC-1 AC210S, Sartorius; precision 0.1 mg). Transpiration rate (J in g m^{-2} s^{-1}) was calculated from the change in fresh weight (ΔFW in g) with time (t in s) divided by the dual projected leaf area (A in m^2).

$$J = \frac{\Delta FW}{\Delta t \times A} \tag{2}$$

The water vapour conductance (g in m s^{-1}) was calculated from the transpiration rate (J) divided by the driving force for water loss. The driving force corresponds to the difference between the concentrations of water vapour (g m^{-3}) in the leaf ($C_{wv\,leaf}$) and the surrounding atmosphere ($C_{wv\,air}$):

$$g = \frac{J}{C_{wv\,leaf} - C_{wv\,air}} = \frac{J}{\alpha_{leaf} \times C_{wv\,sat\,leaf} - \alpha_{air} \times C_{wv\,sat\,air}} \tag{3}$$

The water activity of the air (α_{air}) and the air-water vapour concentration ($C_{wv\,sat\,air}$) over silica gel are nearly zero (Slavík, 1974). The water vapour concentration in the leaf corresponds to the product of water activity in the leaf (α_{leaf}), which is assumed to be unity and the concentration of water vapour saturation at leaf temperature ($C_{wv\,sat\,leaf}$). Leaf temperature was measured using an infrared laser thermometer (Harbor Freight Tools), and the corresponding water vapour saturation concentrations at leaf temperature were derived (Nobel, 2009).

Leaf conductance at a given dehydration point was plotted versus the equivalent RWD. The conductance was high at low RWD and declined with progressive leaf desiccation until reaching a plateau where the conductance was insensitive to a subsequent decline of the RWD (Fig. 2). The constant low conductance corresponds to the minimum leaf conductance and results of the maximum stomatal closure. We determined the RWD at the stomatal closure point (RWD_{SC}) from the leaf drying curves as the mean value of the last conductance before the curve reaches a plateau and the first one on the plateau.

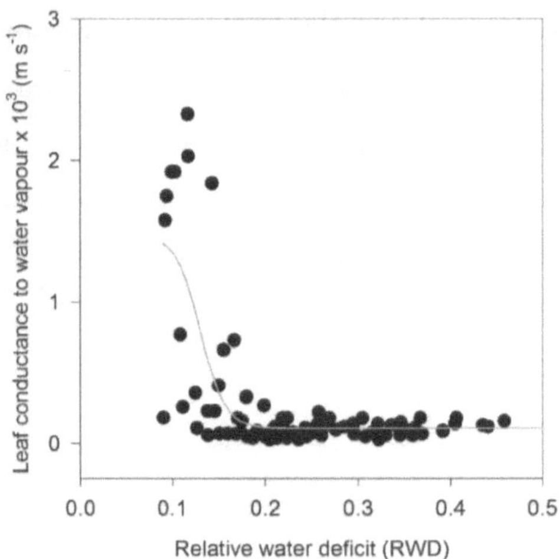

Fig. 2. Leaf conductance as a function of the relative water deficit (RWD) of the Mediterranean *Quercus coccifera*. Each point represents an individual measurement obtained from leaf drying curves of nine leaves at 25°C. A sigmoidal four-parameter curve is fitted to guide the eye. The transition between the declining stage and the plateau stage of leaf conductances represents the stomata closure. After maximum stomata closure, leaf conductances remain constant characterising the minimum leaf conductance phase.

Chemical composition of cuticular waxes

Except for *P. dactylifera*, the cuticular waxes were extracted by dipping the whole leaves (excluding the wounds of cut petioles) twice into 10 mL chloroform ($\geq$ 99.8%, Roth) for 1 min at room temperature. Leaflets of *P. dactylifera* were immersed (apart from cut edges) twice in 25 mL chloroform ($\geq$ 99.8%, Roth) and maintained for 5 min in an ultrasonic bath at room temperature. The cuticular wax amount did not increase with additional extraction steps. *N*-tetracosane (C_{24}; $\geq$ 99.5%, Sigma-Aldrich) was added as an internal standard, and the solutions were reduced to dryness under a gentle flow of nitrogen.

Dry wax samples were derivatised for gas chromatography with *N,O*-bis(trimethylsilyl)trifluoroacetamide (BSTFA, Marchery-Nagel) in dry pyridine ($\geq$ 99.5%, Roth) for 30 min at 70°C. Quantification of cuticular wax components was performed with a gas chromatograph equipped with a flame ionisation detector and an on-column injector (7890A, Agilent Technologies). Separation of compounds was carried out on a fused-silica capillary column (DB1-ms, 30 m length × 0.32 mm ID, 0.1 µm film, Agilent Technologies) with hydrogen as a carrier gas. The temperature program consisted of injection at 50°C for 2 min, raised by 40°C min^{-1} to 200°C, held at 200°C for 2 min, and then raised by 3°C min^{-1} to 320°C, and held at 320°C for 30 min. Single compounds were quantified against the internal standard. Qualitative analysis was carried out using a gas chromatograph equipped with a mass spectrometric detector (5975 iMSD, Agilent Technologies) following the same gas chromatographic conditions, except that helium was the carrier gas.

The weighted average chain-lengths (ACL) of the cuticular waxes of all the 14 plant species were calculated from the chain-length (x_i) and the mass fraction (w_i) of the component (i). The ACL was calculated based on the wax amount coverage (µg cm^{-2}).

$$ACL = \frac{\sum_i x_i \cdot w_i}{\sum_i w_i} \tag{4}$$

In chapter two, weighted median chain-lengths (MCL) for the cuticular waxes of *P. dactylifera* leaflets and *C. colocynthis* leaves were calculated to characterize the chain-length distributions of VLC compounds. Molar coverages for each individual component were calculated from the gas-chromatography data and summed up according to carbon chain-lengths. For each chain-length N_i the mol fraction w_i was obtained and used as weight for further calculating the MCL. For n distinct ordered chain-lengths N_1, N_2 ..., N_n with weights w_1, w_2, ..., w_n, the MCL is the chain-length N_k satisfying

$$\sum_{i=1}^{k-1} w_i \leq \tfrac{1}{2} \text{ and } \sum_{i=k+1}^{n} w_i \leq \tfrac{1}{2} \tag{5}$$

Chemical composition of the cutin matrix

For cutin depolymerisation, delipidated leaves of *C. colocynthis* and dewaxed, enzymatically isolated cuticular membranes of *P. dactylifera* leaflets were transesterified with boron trifluoride in methanol (Fluka) at 70°C overnight to release methyl esters of cutin acid monomers. Sodium chloride-saturated aqueous solution (AppliChem), chloroform and n-dotriacontane (C_{32}; Sigma-Aldrich) as an internal standard were added to all reaction mixtures. From this two-phase system, the de-esterified cutin components were extracted three times with chloroform. The combined organic phases were dried over anhydrous sodium sulfate (AppliChem). All extracts were filtered, and the organic solvent was evaporated under a continuous flow of nitrogen.

Derivatization and subsequent gas chromatographic analysis with mass spectrometric detection were performed according to Leide *et al.* (2011). Separation of cutin mixtures was carried out at 50 kPa for 60 min, 10 kPa min^{-1} to 150 kPa and at 150 kPa for 30 min using a temperature program of 50°C for 1 min, raised by 10°C min^{-1} to 150°C, held at 150°C for 2 min, and then raised by 3°C min^{-1} to 320°C and held at 320°C for 30 min. Quantitative composition of the mixtures was studied using capillary gas chromatography and flame ionisation detection under the same gas chromatographic conditions.

Leaf photosynthetic thermal tolerance

Leaf photosynthetic thermal tolerance (P_{TT}) was determined using the temperature dependent decline in the photochemical efficiency of photosystem II (PSII) according to Knight and Ackerly (2003). We accessed the photochemical efficiency of PSII using the maximum quantum yield of PS II ($F_v\,F_m^{-1}$) followed by an actinic light pulse with a pulse-amplitude modulated fluorometer (Junior PAM, Walz) in dark-adapted leaves. The ratio of variable and maximum fluorescence corresponds to $F_v\,F_m^{-1}$. Leaves were submerged in a temperature controlled water bath rising the temperature at 1°C min^{-1}, and the $F_v\,F_m^{-1}$ was measured every 2.5 minutes. Every single leaf was exposed to the whole temperature range. We determined the critical temperature (T_{crit}) for accessing the leaf photosynthetic thermal tolerance. T_{crit} indicates the high-temperature onset at which PSII damage begins. $F_v\,F_m^{-1}$ was plotted in function of the temperature and, T_{crit} was determined from regression lines fitted to the two linear portions of the plot. The intersection point of both regression lines represents the T_{crit} (Fig. 3).

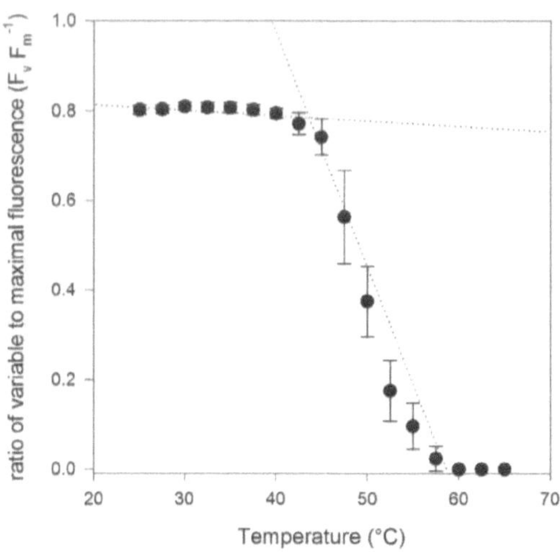

Fig. 3. Temperature effect on the maximum quantum yield of the photosystem II ($F_v\,F_m^{-1}$) of dark-adapted leaves of the tropical dry forest plant *Ceratonia siliqua* (mean ± standard deviation, n = 10).

Statistical analysis

Statistical analyses were performed using the SPSS Statistics software version 23.0 (IBM Corporation). Data were tested for normality by Shapiro-Wilk test. Comparisons between two independent sampling groups were performed using Student's T-Test for normal distributed data and Mann-Whitney U Test was used for those non-normal distributed. Comparisons among more than two independent sampling groups were performed using Analysis of Variance (ANOVA) for normal distributed data followed by a posthoc test satisfying: 1. Tukey test for equal sample sizes and equal variance, 2. Gabriel test for slightly different sample sizes and equal variances, 3. Hochbergs GT2 test for very different sample sizes and equal variances, or 4. Games-Howell for different variances. Those non-normal distributed data were compared using Kruskal-Wallis ANOVA, and subsequent multiple comparisons stepwise step-down. Correlation analyses between two normal distributed variables were performed using Pearson correlation and Spearman Rank correlation was used for those non-normal distributed variables. In all cases the level of significance was $p < 0.05$.

Chapter 1

Cuticular leaf wax deposition and cuticular transpiration barrier in *Quercus coccifera* L.: does the environment matter?

Abstract: Plants prevent uncontrolled water loss by synthesising, depositing, and maintaining a hydrophobic layer (the plant cuticle) over their primary aerial organ surfaces. *Quercus coccifera* L. can plastically respond to environmental conditions at cuticular level. Plants grown under Mediterranean-type climate (MED) accumulate leaf cuticular waxes over the stomata, reducing the stomatal conductance. However, the waxes contribution to reducing stomatal conductance under Temperate-type climate (TEM) is small or absent. Despite the ecophysiological importance of the cuticular waxes of *Q. coccifera*, the chemical wax composition and its contribution to avoid uncontrolled dehydration under MED and TEM climates remain unknown. Thus, we investigated (quantitatively and qualitatively) the cuticular wax composition by gas-chromatography and determined the minimum leaf conductance (g_{min}) of *Q. coccifera* from both MED and TEM climates. The leaf cuticular wax deposition increased by 2.6-fold under MED climate and contained higher amounts of all identified compound classes (except for the equivalence of alkanoic acids). The total very-long-chain aliphatics and cyclic compounds were 3.5-fold and 2.2-fold higher than under TEM climate, respectively. Although MED leaves presented higher wax load, the g_{min} remained unaltered between habitats. Our findings suggest that the genetic control on the biosynthesis of equivalent chemical functional groups might compensate the higher wax deposition of *Q. coccifera*, thereby contributing equally to the cuticular transpirational barrier. In conclusion, we showed that the cuticular wax deposition per unit area of *Q. coccifera* is highly affected by environmental features, whereas genetic factors might mainly determine its cuticular water permeability.

Keywords: cuticular waxes, Mediterranean climate, Temperate climate, cuticular conductance, leaf area reduction, stomata encryption.

1.1 Introduction

The first plant transition from an exclusively aquatic to a terrestrial environment happened approximately 440 million years ago (Niklas, 1997 and 2016; Lewis and McCourt, 2004). Together with the new environment came up a set of challenges, including desiccation, gravity, and increased temperatures and UV radiation (Waters, 2003; Leliaert *et al.*, 2011, Yeats and Rose, 2013). Since then, plants have evolved a multitude of morphological and physiological features that allow them to cope with these new challenges. However, the capacity to synthesise, deposit and maintain a hydrophobic surface layer, named cuticle, onto the surface of aerial organs has been claimed to be the most critical adaptative trait for plant survival in the highly dehydrating terrestrial habitat (Yeats and Rose, 2013).

The primary function of plant cuticle is avoiding uncontrolled water loss. Plant cuticle consists of a cutin matrix intermeshed and coated with waxes. The cutin matrix is mainly composed of C_{16} to C_{18} hydroxy alkanoic acids and their derivatives (Pollard *et al.*, 2008), whereas typical waxes comprise very-long-chain aliphatics and cyclic molecules (Jetter *et al.*, 2006). Waxes mainly constitute the cuticular barrier against the water diffusion through the cuticle (Schönherr, 1976; Burghardt and Riederer, 2006). Besides avoiding dehydration, plants also have to acquire sufficient carbon dioxide for photosynthesis. Here appears a classic plant trade-off, which involves stomatal transpiration and photosynthesis. While the stomata are open to CO_2 uptake, plants inevitably lose water to the surrounding atmosphere. Thus, plants must optimise CO_2 uptake and water loss to reach optimal performance, particularly under arid environments.

Drought is the major limiting factor for production, growth, development, and distribution of plants in Mediterranean environments. *Quercus coccifera* L. is a sclerophyllous evergreen shrub which can withstand prolonged drought periods under high temperatures and low air humidity in the Mediterranean region (Vilagrosa *et al.*, 2003, Peguero-Pina *et al.*, 2008). This species is one of the most representative elements composing the shrub-land flora in the aridest regions of the Iberian Peninsula (Peguero-Pina *et al.*, 2008), but its distribution even reaches temperate climatic conditions in the Iberian Atlantic coast (Castro Díez and Navarro, 2007). *Q. coccifera* is capable of plastically respond to environment variations, allowing this species to occur in these contrasting habitats (Rubio de Casas *et al.*, 2007). Roth-Nebelsick *et al.* (2013) demonstrated that *Q. coccifera* is capable of developing leaf epidermal structures to reduce the stomatal conductance when growing in the aridest Mediterranean-subtype climate. Cuticular waxes form chimney-like structures surrounding the stomata, reducing the effective stomata pore opening from 32 to 5 μm^2. Moreover, Peguero-Pina *et al.* (2016) showed that under Temperate climate the cuticular waxes of *Q. coccifera* leaves do not reduce the stomatal pore as much as in Mediterranean climate. The authors attributed this fact to the plasticity of stomatal protection, at a cuticular level, in response to contrasting climatic conditions.

Wax chimney-like structures protecting the stomata have been reported to many plant species (Scareli-Santos *et al.*, 2013; Panahi *et al.*, 2012; Müller *et al.*, 2017) and have been considered as an adaptation to arid environments (Raven *et al.*, 2005). Despite the physiological and ecological importance of the cuticular waxes, either on the cuticular transpiration barrier or the stomatal protection, the relationship between the chemical wax composition and the cuticular water permeability in *Q. coccifera* remains unknown.

This study aims to investigate the environment effect on the leaf wax deposition and the efficiency of the cuticular transpiration barrier of *Q. coccifera*. We hypothesised that in Mediterranean plants when compared with Temperate plants: (i) the amount and/or composition of cuticular waxes widely differ and (ii) the cuticle is equally efficient to avoid water loss. From these hypotheses, we predicted that in comparison to temperate plants, Mediterranean plants (i) possess a high degree of stomatal encryption due to higher cuticular wax deposition (majorly influenced by environment) per unit area and (ii) the cuticular transpiration barrier (mostly believed to be determined by genetic factors) remains unaltered. We tested Hypothesis (i) by qualitatively and quantitatively determining the composition of the leaf cuticular waxes and (ii) by measuring the minimum leaf conductance at 25°C for plants from both the Mediterranean and the Temperate sites.

1.2 Results

1.2.1 Leaf traits

Leaf traits of *Q. coccifera* were calculated (Fig. 4). The mean value (n ≥ 16) of leaf mass per area (LMA) in the MED site amounted to 245.7 ± 14.4 g m^{-2}, which was statistically different (P < 0.05) from the mean value in the TEM site (152.0 ± 7.1 g m^{-2}). Leaf water content (LWC) was statistically lower (p < 0.05) in MED plants (0.42 ± 0.02 g g^{-1}) than that of TEM plants (0.48 ± 0.01 g g^{-1}). Leaf dry weight (DW) did not show statistically significant differences at p < 0.05 between sites (0.02 ± 0.01 and 0.03 ± 0.01 g for MED and TEM sites, respectively). Last, leaf size, accessed as dual leaf projected area (LA), was reduced significantly (p < 0.05) in the MED site (0.18 ± 0.05 · 10^{-3} m^2) when compared to the TEM site (0.37 ± 0.16 · 10^{-3} m^2).

1.2.2 Leaf surface properties

Samples of one-year-old leaves of *Q. coccifera* were analysed with scanning electron microscopy to examine the morphology of the leaf surface. Presence of trichomes, stomata and cuticular wax structures were the primary investigated features. *Q. coccifera* leaves from TEM and MED sites had few trichomes on both adaxial and abaxial leaf surfaces and possessed stomata only on the abaxial surface. The stomata distribution occurs without any distinct pattern along the leaf surface. The cuticular surface structure showed a continuous cuticular wax layer with the presence of few wax granules. On the abaxial leaf surface of MED plants, the wax granules were more abundant, and the continuous cuticular wax layer was projected over the stomata reducing the stomatal pore when compared to TEM plants (Fig. 5), as reported by Roth-Nebelsick *et al.* (2013).

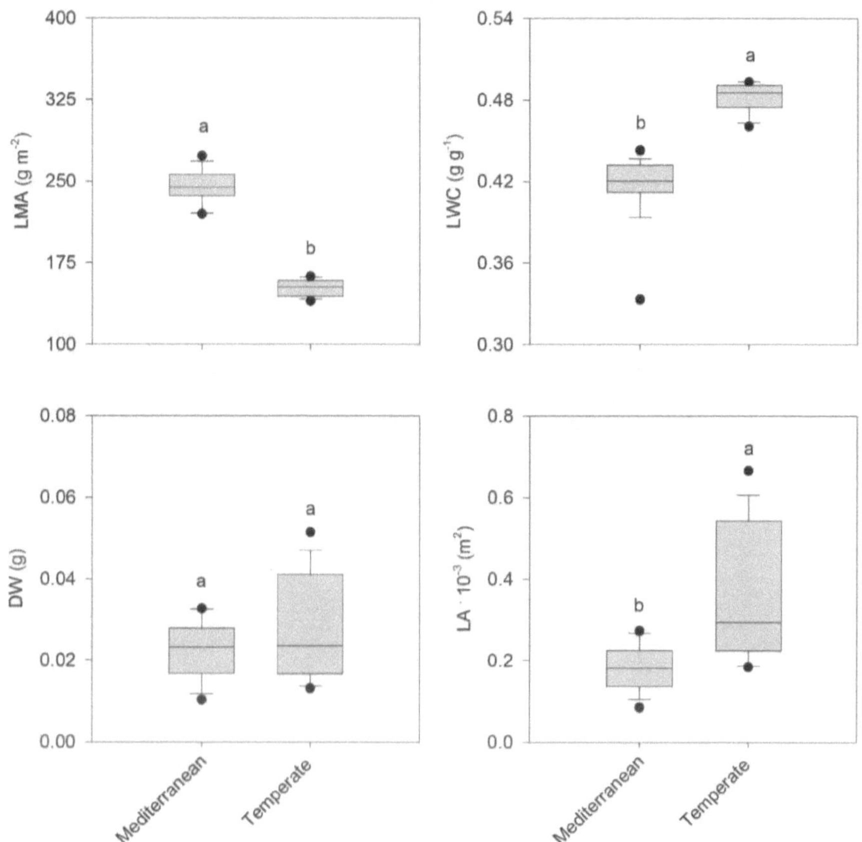

Fig. 4. Leaf traits of *Quercus coccifera* from Mediterranean (MED) and Temperate (TEM) sites (n ≥ 16). Plants growing under MED conditions presented higher leaf mass per area (LMA) than under TEM conditions. Moreover, MED plants possessed lower leaf water content (LWC) and had the dual projected leaf area (LA) reduced in comparison to TEM plants. Leaf dry weight (DW) did not change between habitats. Different letters indicate significant differences at $P < 0.05$ between the Mediterranean and Temperate sites.

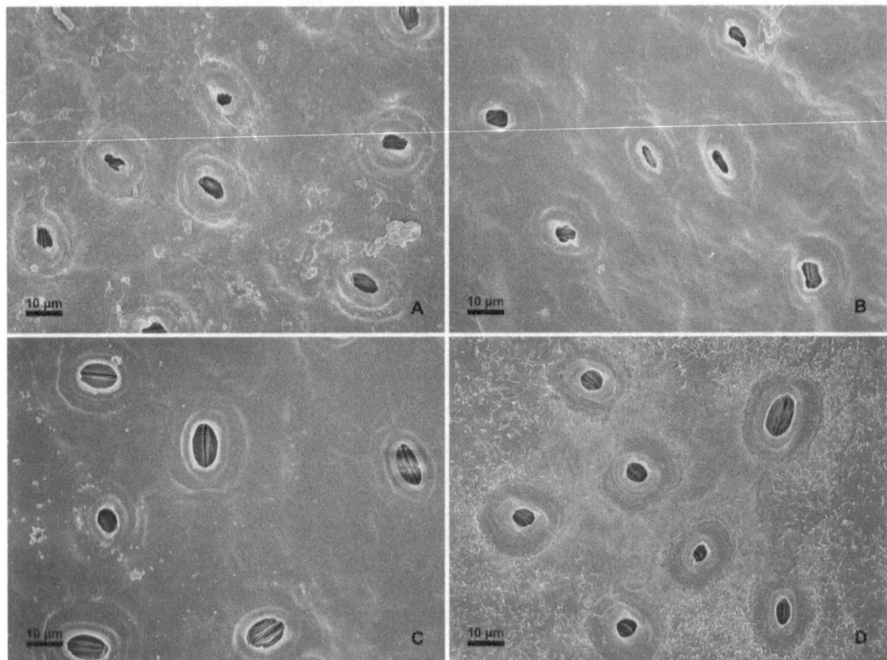

Fig. 5. Scanning electron microscopy of abaxial surface of *Quercus coccifera* leaves. The cuticular wax layer of untreated leaves extends over the stomata reducing stomata opening in MED site (A) in comparison to TEM site (B). After wax extraction, the difference in stomatal pore dimension between MED site (C) and TEM site (D) disappears. See Roth-Nebelsick *et al.* (2013) for details.

1.2.3 Leaf minimum conductance

Leaf minimum conductances (g_{min}) with maximal stomal closure were determined at 25°C from drying curves of fully developed leaves (n ≥ 16) to identify possible environmental influence on the cuticular transpiration barrier efficiency of 8-year-old *Q. coccifera* plants. The first stage of leaf drying curves was characterised by high conductances that decrease with leaf dehydration until reaching a plateau of constant conductance values when stomata maximally close. G_{min} mean values were 11.96 ± 3.71 and 12.43 ± 4.15 x 10^{-5} m s^{-1} for MED and TEM sites, respectively (Fig. 6). There was not statistically significant differences between habitats (p < 0.05).

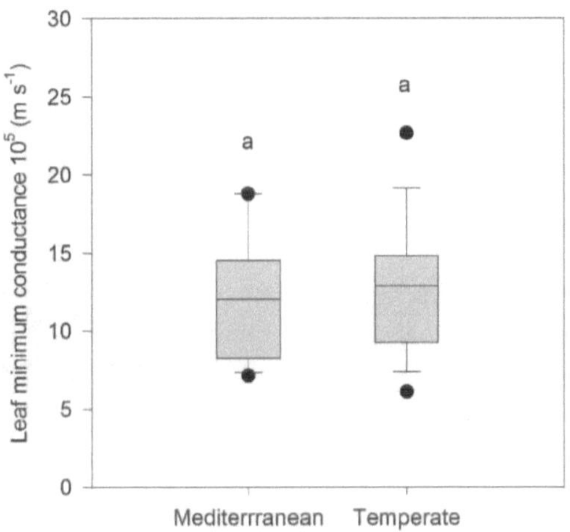

Fig. 6. Minimum conductance (g_{min}) of one-year-old leaves of *Quercus coccifera* obtained from drying curves at 25°C. The g_{min} of 8-year-old plants did not differ between Mediterranean (MED) and Temperate (TEM) climate conditions (t (31) = -0.332, p = 0.742).

1.2.4 Chemical composition of leaf cuticular waxes

The cuticular waxes *Q. coccifera* were analysed qualitatively and quantitatively to investigate the potential effect of environment on the leaf wax deposition. More than 84% of the wax extract was identified using GC-MS. The cuticular wax coverage of one-year-old leaves was 2.6-fold higher for plants growing in MED site than that of TEM site (Table 2). Triterpenoids were the most abundant class of wax constituents (49.2 and 60.5% of total wax in MED and TEM site, respectively), and germanicol (28.9 and 26.1%) and lupeol (8.2 and 12.1%) were the major constituents. Very-long-chain aliphatic compounds with chain lengths ranging from C_{20} to C_{51} corresponded to 34.7 and 26.2% of the total wax in TEM and MED sites, respectively. Within the aliphatic fraction, alkane was the main compound class (10.1 and 8.0%). The alkane fraction comprised a homologous series of *n*-alkanes from C_{25} to C_{32} with odd-numbered *n*-alkanes dominating above even numbered. Nonacosane was the major constituent of the aliphatic fraction (4.6 and 3.2%) followed by hentriacontane (2.8 and 2.8%). The weighted average chain-length of the very-long-chain aliphatic wax components was 32.4 and 32.9 for MED and TEM sites, respectively.

Table 2. The cuticular leaf waxes of *Quercus coccifera* are mainly composed of cyclic aliphatic compounds, regardless of the habitat. Each value represents the mean ± standard deviation (*n* = 4 biological replicates).

Compound class	Carbon chain length	Coverage (μg cm^{-2})	
		Temperate site	Mediterranean site
n-alkanes	25	0.022 ± 0.006	0.073 ± 0.006
	26	0.009 ± 0.001	0.064 ± 0.008
	27	0.048 ± 0.020	0.200 ± 0.046
	28	0.028 ± 0.024	0.198 ± 0.021
	29	0.426 ± 0.134	1.611 ± 0.748
	30	0.114 ± 0.050	0.208 ± 0.016
	31	0.371 ± 0.040	0.980 ± 0.222
	32	0.039 ± 0.019	0.209 ± 0.062
primary alcohols	22	0.005 ± 0.002	0.013 ± 0.005
	23	0.003 ± 0.001	0.016 ± 0.005
	24	0.024 ± 0.006	0.136 ± 0.064
	25	0.010 ± 0.004	0.029 ± 0.013
	26	0.075 ± 0.019	0.102 ± 0.014
	27	0.014 ± 0.007	0.040 ± 0.018
	28	0.023 ± 0.016	0.073 ± 0.011
	30	0.101 ± 0.042	0.541 ± 0.267
	31	0.092 ± 0.067	0.376 ± 0.085
	32	0.191 ± 0.119	1.011 ± 0.557
	33	0.052 ± 0.030	0.220 ± 0.015
	34	0.058 ± 0.043	0.197 ± 0.019
alkanol cetates	26	-	0.025 ± 0.014
	27	-	0.027 ± 0.016
	28	-	0.135 ± 0.114
	29	0.013 ± 0.010	0.062 ± 0.026
	30	0.026 ± 0.015	0.426 ± 0.245
	31	-	0.108 ± 0.035
alkanals	28	0.016 ± 0.007	0.035 ± 0.020
	30	0.085 ± 0.020	0.202 ± 0.085
	32	0.066 ± 0.038	0.315 ± 0.021
alkanoic acids	20	0.006 ± 0.002	0.025 ± 0.008
	21	0.002 ± 0.001	0.011 ± 0.002
	22	0.004 ± 0.002	0.077 ± 0.074

Compound class	Carbon chain length	Coverage (µg cm^{-2})	
		Temperate site	Mediterranean site
	23	0.004 ± 0.002	0.032 ± 0.006
	24	0.018 ± 0.008	0.123 ± 0.099
	25	0.007 ± 0.004	0.037 ± 0.027
	26	0.026 ± 0.023	0.111 ± 0.075
	27	0.023 ± 0.007	0.044 ± 0.021
	28	0.102 ± 0.029	0.169 ± 0.122
	29	0.102 ± 0.071	0.181 ± 0.075
	30	0.247 ± 0.136	0.848 ± 0.550
	31	0.103 ± 0.010	0.220 ± 0.064
	32	0.084 ± 0.064	0.521 ± 0.466
	33	0.026 ± 0.012	0.081 ± 0.027
	40	0.020 ± 0.007	0.110 ± 0.026
	42	0.035 ± 0.013	0.189 ± 0.046
	44	0.109 ± 0.026	0.467 ± 0.120
	45	0.018 ± 0.001	0.075 ± 0.014
alkyl esters	46	0.152 ± 0.031	0.452 ± 0.078
	47	0.023 ± 0.004	0.069 ± 0.021
	48	0.124 ± 0.032	0.306 ± 0.040
	49	0.020 ± 0.006	0.050 ± 0.014
	50	0.091 ± 0.025	0.169 ± 0.014
	51	0.024 ± 0.004	0.039 ± 0.011
glycerol esters	16	0.087 ± 0.016	0.059 ± 0.011
	18	0.092 ± 0.008	0.049 ± 0.009
total very-long-chain aliphatics		*3.461 ± 0.619*	*12.144 ± 2.088*
alpha-amyrin		0.169 ± 0.049	0.244 ± 0.088
beta-amyrin		0.287 ± 0.073	0.985 ± 0.097
beta-sitosterol		1.134 ± 0.805	0.236 ± 0.063
betulin		0.198 ± 0.068	0.339 ± 0.050
betulinic acid		0.260 ± 0.200	0.591 ± 0.148
erythrodiol		0.106 ± 0.025	-
fridelin		0.048 ± 0.045	0.232 ± 0.084
fridelinol		0.112 ± 0.054	0.317 ± 0.166
germanicol		3.448 ± 0.550	10.105 ± 2.279
germanicone		0.031 ± 0.011	0.127 ± 0.012
lupeol		1.601 ± 0.364	2.881 ± 1.388
oleanolic acid		0.171 ± 0.099	0.232 ± 0.038
triterpenoid 1		-	0.059 ± 0.020
triterpenoid 2		-	0.268 ± 0.072
ursolic acid		0.082 ± 0.053	0.208 ± 0.083
uvaol		0.208 ± 0.069	0.276 ± 0.035
total cyclic aliphatics		*7.856 ± 1.674*	*17.101 ± 3.032*
beta-tocopherol		0.113 ± 0.084	0.075 ± 0.033
delta-tocopherol		0.027 ± 0.024	0.043 ± 0.007
total cyclic aromatics		*0.140 ± 0.094*	*0.117 ± 0.040*

Compound class	Carbon chain length	Coverage ($\mu g\ cm^{-2}$)	
		Temperate site	Mediterranean site
not identified		1.751 ± 0.549	5.612 ± 2.049
total		*13.208 ± 2.748*	*34.974 ± 6.636*

The cuticular wax composition changed in response to the environment. Besides possessing an increased leaf wax coverage, plants growing under Mediterranean conditions had 2.2-fold more triterpenoids and 3.5-fold more aliphatic compounds than those growing under Temperate conditions. Within the aliphatic fraction, except for alkanoic acids, all the compound classes increased ($p < 0.05$) in function of the growing habitat (Fig. 7).

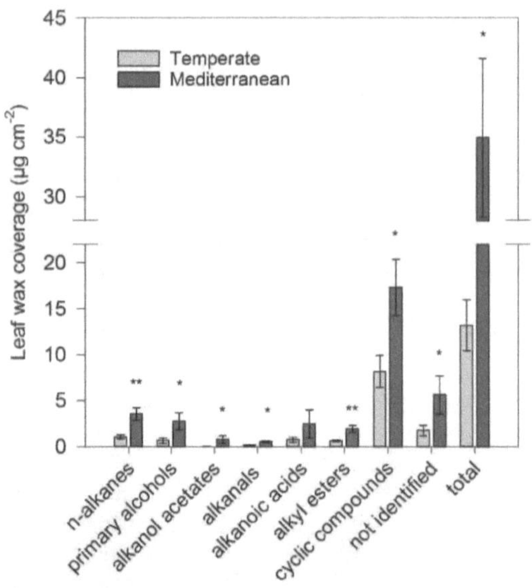

Fig. 7. Cuticular wax coverage of *Quercus coccifera* leaves arranged by compound class. Each value represents the mean ± standard deviation ($n = 4$). Asterisk indicates significant difference at $p < 0.05$ and double-asterisk at $p \leq 0.01$ between the Temperate (TEM) and the Mediterranean (MED) sites.

1.3 Discussion

Drought often goes along with high temperatures during summer time in Mediterranean-type climates. Leaf size plays an important role in water and leaf energy balance, especially in arid environments. Also, leaf mass per area (LMA) reflects the intrinsic relation between carbon gain and longevity (Díaz *et al.,* 2016), while leaf water content (LWC) roughly indicates leaf density (Garnier and Laurent, 1994). *Q. coccifera* responds to Mediterranean-type climate reducing leaf size (2-fold in comparison with Temperate-type climate) and LWC (1.1-fold), and increasing LMA (1.6-fold). Hence, the small leaves with low LWC and high LMA in Mediterranean site may be associated with high heat and drought resistances of *Q. coccifera.*

Along the summer days, it is common in Mediterranean sites that the temperature rises while the air humidity reaches low values, resulting in an intensified driving force to water loss by transpiration. Although *Q. coccifera* is well adapted to arid conditions, this species also occurs in the Iberian Atlantic coast under Temperate conditions. The milder temperatures and the higher air humidity in Temperate-type climate reduce the evaporative demand of plants by reducing vapour pressure deficit (VPD). Hence, one may expect the minimum leaf conductance (g_{min}) of *Q. coccifera* to be lower in Temperate site due to the reduced VPD between leaf and atmosphere, and higher in the Mediterranean site in response to an intensified driving force to water loss. However, our findings do not support this idea since the g_{min} of *Q. coccifera* remained unaffected by these contrasting climatic conditions.

Water permeability of astomatous isolated cuticles of *Citrus aurantium* grown in environmental chambers at temperatures ranging from 15 to 35°C and humidities of 50 and 90% remained unaltered, regardless of environment (Geyer and Schönherr, 1990). Also, permeances of isolated cuticles from 21 plant species tended to cluster according to life forms and climate of origin (Schreiber and Riederer, 1996). In line with our hypothesis, the authors suggested that genetic factors primarily determine the water permeabilities of plant cuticles, once the investigated leaves grew under similar climatic conditions. Riederer and Schreiber (2001) reported lower cuticular permeances for xeromorphic plants typical from the Mediterranean climate in comparison with deciduous plants characteristics from Temperate climate. Based on this finding, intuitively, one might assume that plant species growing under arid conditions possess comparatively low cuticular water permeabilities. Schuster *et al.* (2017) tested this assumption analysing 382 measurements of cuticular permeabilities from 160 plant species extracted from the literature. The plant species were assigned to 11 life form groups and only in two particular cases, epiphytes and climbers/lianas, the cuticular permeabilities were unusually low. At a congeneric level, Gil-Pelegrín *et al.* (2017) investigated 11 *Quercus* species growing in a common garden under Mediterranean-type climate and found out that g_{min} of evergreen Mediterranean oaks slightly differs from that of deciduous Temperate oaks. However, the g_{min} of Mediterranean and Temperate deciduous species remained unaltered. It supports our results, suggesting that cuticular water permeability is a species-specific feature, majorly determined by genetic factors and plant life strategy, suffering minor or absent influence from the environment.

Cuticular waxes compose the primary barrier against uncontrolled water loss and vary in response to biotic (e.g. organs and development stages) and abiotic factors (e.g. drought and relative humidity). Leaf wax deposition per unit area of *Q. coccifera* increased under Mediterranean conditions, corroborating our hypothesis that MED plants possess a differentiated amount of cuticular waxes in comparison with TEM plants. The leaf wax coverage of Mediterranean plants was 2.6-fold higher than that of Temperate plants. In line with our findings, the leaf cuticular wax of tree tobacco (*Nicotiana glauca*) increased up to 2.5-fold after induced drought stress (Cameron *et al.*, 2006). Adittionaly, Kim *et al.* (2007) imposed drought on 18 sesame (*Sesamum indicum*) cultivars by withholding irrigation, which caused an increase of leaf wax amount on 15 cultivars. In the same direction, Kosma *et al.* (2009) reported that the model plant *Arabidopsis*, when exposed to a water deficit treatment, exhibited an increase of circa 75% in the total leaf wax amount. Le Provost *et al.* (2013) analysed the effect of rainfall exclusion on the needle wax deposition of maritime pine (*Pinus pinaster*) and showed that the wax content increased in drought-stressed plants.

Q. coccifera from Temperate-type climate experienced during summer (the aridest season) over 60% of the daily hours with non-limiting light conditions registered relative humidity (RH) values above 70%, whereas under Mediterranean-type climate almost 50% of the daily hours registered RH below 30% (Roth-Nebelsick *et al.*, 2013). High RH tends to reduce the evaporative demand by decreasing the vapour pressure deficit (VPD). Plants under low VPD transpire less, improving their water status and reducing the demand for cuticular waxes (Lihavainen *et al.*, 2017). Adittionally, Koch *et al.* (2006) showed that the wax content of *Brassica oleracea* strongly declines in response to low VPD. Thus, the low VPD values also support the reduced leaf wax deposition of *Q. coccifera* under Temperate-type climate.

Triterpenoids mainly comprised (> 49% of the total amount) the cuticular leaf waxes of *Q. coccifera*, regardless of habitat. Germanicol, the principal wax constituent, has been reported in the cuticular waxes of tomato fruit (*Lycopersicon esculentum*, Bauer *et al.*, 2004), petals and leaves of heather (*Calluna vulgaris*, Szakiel *et al.*, 2013), and grape berries (*Vitis vinifera*, Pensec *et al.*, 2014). Triterpenoids have been associated with protection against herbivory (Reichardt *et al.*, 1984; Oliveira and Salatino, 2000) and with the stabilisation of heat-stressed cuticles (Schuster *et al.*, 2016). However, the contribution of triterpenoids to avoid uncontrolled water loss has been considered small or absent (Leide *et al.*, 2007 and 2011; Buschhaus and Jetter, 2012; Schuster *et al.*, 2016, Jetter and Riederer, 2016). Thereby, the efficacy of the transpirational barrier has been attributed to the very-long-chain aliphatics compounds located in both the intra and epicuticular waxes (Jetter and Riederer, 2016). These findings are in line with the molecular structure model of cuticular waxes proposed by Riederer and Schreiber (1995). According to the authors, the cuticular waxes are multiphase systems made up of mobile amorphous zones within highly structured crystalline domains. This model predicts that the very-long-chain aliphatic compounds build up the impermeable crystalline domains, and the amorphous zones incorporate the chain ends and cyclic molecules. Hence, the model assumes that the very-long-chain aliphatics constitute the cuticular transpirational barrier in plants.

Q. coccifera had 3.5-fold more very-long-chain aliphatic compounds in the Mediterranean site than in Temperate site, and one may be inclined to expect that high amount of very-long-chain aliphatics should reduce the g_{min} in Mediterranean site. However, cuticular water permeability is not related to the amount of total cuticular waxes or cuticle thickness (Schreiber and Riederer, 1996; Riederer and Schreiber, 2001). Although plants living in both Mediterranean and Temperate sites quantitatively present distinct amounts of compound classes and total wax load, the qualitative contribution of each class and the carbon chain length of homologous compounds are very similar. Thus, the genetic control on the biosynthesis of equivalent chemical functional groups might compensate the quantitative plasticity of the cuticular waxes deposition of *Q. coccifera*, thereby contributing equally to the cuticular transpirational barrier.

In conclusion, we showed that the leaf cuticular wax deposition per unit area of *Q. coccifera* increases under Mediterranean-type climate, but the cuticular water permeability remains unaltered regardless of habitat. It suggests that genetic factors mainly control the cuticular water permeability whereas environmental features, especially drought and air humidity, profoundly affect the leaf wax deposition. We also suggest that the high wax deposition per unit area associated with the leaf size reduction is crucial for stomata protection of *Q. coccifera* under Mediterranean-type climate, reducing stomatal conductance and improving the plant water status as reported in previous studies.

Chapter 2

Comparative analysis of temperature effects on the cuticular transpiration barrier of the desert water-spender *Citrullus colocynthis* and the water-saver *Phoenix dactylifera**

*In memory of Otto Ludwig Lange (1927 – 2017).

Abstract: In deserts, drought often coincides with high temperatures, resulting in an exacerbated driving force to water loss. Hence, an efficient water loss barrier to the hot and dry atmosphere is crucial for ensuring plant survival. This work aims to elucidate whether specific cuticular transpiration barrier properties of desert plants coincide with contrasting life strategies to cope with high temperatures and dry environments. We hypothesize that in a water-saver plant (*Phoenix dactylifera*) when compared to a water-spender plant (*Citrullus colocynthis*): (i) the cuticle is more efficient to avoid water loss, (ii) the cuticle is more stable under thermal stress, and (iii) the amount and/or composition of cuticular compounds widely differ. The minimum leaf water conductance (g_{min}) of *P. dactylifera* at 25°C was six-fold lower than that of *C. colocynthis*. The temperature effect on the increase of g_{min} from 25°C to 50°C was three-fold lower in *P. dactylifera*. Concerning the cuticular chemical composition, *P. dactylifera* had a higher amount of cutin monomers and a higher C_{18}/C_{16} ratio of cutin acids compared to *C. colocynthis*. Very-long-chain aliphatic compounds were the main cuticular wax components of both plant species (>87%). However, in comparison with *C. colocynthis*, the weighted median carbon chain-length (MCL) of aliphatic components of *P. dactylifera* was high. We proposed that the high concentration of very-long-chain aliphatic compounds with remarkable high MCL might play a key role in the cuticular transpiration barrier and stabilisation of heat-stressed cuticles.

Keywords: desert ecology, drought, thermal stress, cuticular conductance, cuticular waxes

2.1 Introduction

Desert plants have evolved a multitude of adaptations that allow them to cope with aridity and intense solar radiation. Different desert life forms and structural features result from the specialisation to live in arid environments. There are numerous strategies for dealing with drought stress, and plants can be classified along an "adaptation continuum" (resistance) from drought tolerance to drought avoidance (Smith *et al.*, 1997). Drought avoiding plants present two contrasting strategies in response to a low water source (Baquedano and Castillo, 2006). Water-saver plants have strong stomatal control and prevent damage by closing stomata before any change takes place in the leaf water potential. In contrast, water-spender plants decrease the water potential allowing them to extract water from the soil to compensate the high water loss by stomatal transpiration (Guehl and Aussenac, 1987; Lo Gullo and Salleo, 1988).

In hot-deserts, drought often coincides with high temperatures. Along the day, very often the air humidity reaches low values, while the temperature increases, resulting in an exacerbated driving force to water loss by transpiration. Therefore, efficient control of water loss to the hot and dry surrounding atmosphere is of fundamental importance for ensuring plant survival, viability, and reproductive fitness. Stomata and cuticle jointly work to maintain a favourable plant water status. The cuticle, a cutin matrix with cuticular waxes embedded within and deposited on its surface, covers all primary aerial parts of terrestrial higher plants. The cutin matrix primarily consists of cross-linked C_{16} and C_{18} hydroxy and hydroxy epoxy alkanoic acids whereas the characteristic cuticular waxes comprise very-long-chain aliphatic compounds and cyclic components, particularly triterpenoids (Jetter *et al.*, 2006).

Despite its narrow thickness (about 1 µm to 20 µm; Jeffree, 2006), the plant cuticle has essential autoecological functions. Water-stressed plants reduce the water loss closing their stomata, and the water permeability of the cuticle determines the minimum and inevitable water loss. Hence, avoiding uncontrolled dehydration by a low cuticular water permeability is one of the principal features ensuring the establishment and survival of plants in arid environments.

Leaf cuticular permeability varies among plant species from 10^{-7} to 10^{-4} m s^{-1} (Riederer and Schreiber, 2001). Extraction of the cuticular waxes increases the cuticular water permeability by up to several orders of magnitude (Schönherr, 1976; Schönherr and Lendzian, 1981). The influence of cuticle thickness or the amount of cuticular waxes on the cuticular water permeability is small or absent (Schreiber and Riederer, 1996; Riederer and Schreiber, 2001).

The temperature effect on the cuticular water permeability is one of the leading parameters influencing the physiological and ecological role of the cuticle. The cuticular permeability of all non-desert plants studied heretofore have increased slightly at temperatures from 15°C to 35°C, while temperatures above 35°C caused a drastic increase in the permeability (Schreiber, 2001; Riederer, 2006). So far, a single hot-desert plant (*Rhazya stricta*) has been studied in detail. Compared to non-desert plants, the cuticular water permeability of *R. stricta* continuously increased between 15°C and 50°C without an abrupt change in slope (Schuster *et al.*, 2016).

The longstanding successful adaptation of xerophytes to hot-deserts might require different solutions to the same environmental pressures. Pioneer in desert ecological studies, Lange (1959) pointed out contrasting eco-physiological strategies of two desert plants concerning water relation and thermal tolerance. *Citrullus colocynthis* (L.) Schrad. (Cucurbitaceae) is a typical perennial water-spender plant (Lange, 1959), non-hardy drought-resistant with a deep root system (Si *et al.*, 2009). *Phoenix dactylifera* L. (Arecaceae) is a typical woody water-saver crop (Lange, 1959) able to thrive in hot and dry conditions, with low or no rain, as long as there is constant moisture around the roots (Barrow, 1998). *C. colocynthis* exhibits a vine-like life form and is particularly vulnerable to heat. Lange (1959) showed that even under intense solar radiation, an active transpiring leaf maintains its temperature much lower than that of the surrounding air and far away from the limit of heat tolerance of *C. colocynthis*. The cooling effect of transpiration of this plant species avoids death by overheating (Lange, 1959; Althawadi and Grace, 1986). In comparison, a leaflet of *P. dactylifera* exposed to high solar radiation reaches temperatures above that of the air temperature. Despite leaflet overheating, its temperature does not exceed the limit of heat tolerance ensuring the physiological maintenance of *P. dactylifera* (Lange, 1959). Although this information has been available for a long time and eco-physiological differences among desert life forms are well known, only a few studies have dealt with patterns in their cuticular transpiration barrier.

Aiming to close this gap in knowledge, we investigated the minimum leaf water conductance of two desert plants (*C. colocynthis* and *P. dactylifera*) within the range of ecologically relevant temperatures. We characterised the structure of the leaf cuticular surface using scanning electron microscopy and analysed the qualitative and quantitative composition of the cuticular waxes and the cutin matrix of *C. colocynthis* and *P. dactylifera*. These plant species are excellent for comparative studies because they represent two contrasting non-succulent life forms able to deal differently with high temperatures and drought (Lange 1959). Our results are discussed to evaluate our hypotheses that the cuticle of a water-saver plant compared with a water-spender plant (i) is more efficient to avoid water loss, (ii) is more stable under thermal stress, and (iii) the leaf cuticular composition differs widely.

2.2 Results

2.2.1 Morphological and anatomical leaf traits

The morphological and anatomical traits of *Citrullus colocynthis* leaves and *Phoenix dactylifera* leaflets were calculated (Table 3). Leaf mass per area (LMA) of *C. colocynthis* amounted to 60.11 ± 15.95 g m^{-2} (mean $\pm$ standard deviation), and the water content per leaf area (WC LA^{-1}) amounted to 298.84 ± 52.84 g m^{-2}. The dual projected leaf area was $0.45 \pm 0.13 \times 10^{-2}$ m^2, the saturation fresh weight 0.82 ± 0.34 g, and the dry weight 0.14 ± 0.06 g. In *P. dactylifera*, the leaflet mass per area (LMA) amounted to 161.83 ± 22.41 g m^{-2} and the water content per leaf area (WC LA^{-1}) amounted to 186.57 ± 20.17 g m^{-2}. The dual projected leaflet area was 0.74 ± 0.25 m^{-2}, the saturation fresh weight 1.27 ± 0.42 g, and the dry weight 0.59 ± 0.21 g.

Table 3. Morphological and anatomical traits of *Citrullus colocynthis* leaves and *Phoenix dactylifera* leaflets. The dual projected leaf area (LA), the saturation fresh weight (FW$_{sat}$) and the dry weight (DW) were measured. The leaf water content (LWC) and the leaf mass per area (LMA) were calculated, respectively. Each value represents the mean $\pm$ standard deviation ($n \geq 61$).

Leaf traits	*Citrullus colocynthis*	*Phoenix dactylifera*
LA $\times$ 10^{-2} (m^2)	0.45 ± 0.13	0.74 ± 0.25
FW$_{sat}$ (g)	0.82 ± 0.34	1.27 ± 0.42
DW (g)	0.14 ± 0.06	0.59 ± 0.21
LWC (g g^{-1})	0.83 ± 0.02	0.54 ± 0.03
LMA (g m^{-2})	60.11 ± 15.95	161.83 ± 22.41

2.2.2 Leaf surface properties

Samples of full-expanded leaves of *C. colocynthis* and leaflets of *P. dactylifera* were analysed by scanning electron microscopy to examine the morphology of the leaf surface. Presence of stomata, trichomes, and cuticular wax structures were the primary investigated features. Leaves of *C. colocynthis* and leaflets of *P. dactylifera* had stomata on both the adaxial and the abaxial leaf surface. Stomata of *C. colocynthis* presented distribution without any distinct pattern along the leaf surface. The cuticular surface structure showed a continuous cuticular wax layer without explicit crystalloid sculptures and the presence of trichomes varying in density between the adaxial and abaxial surface (Fig. 8).

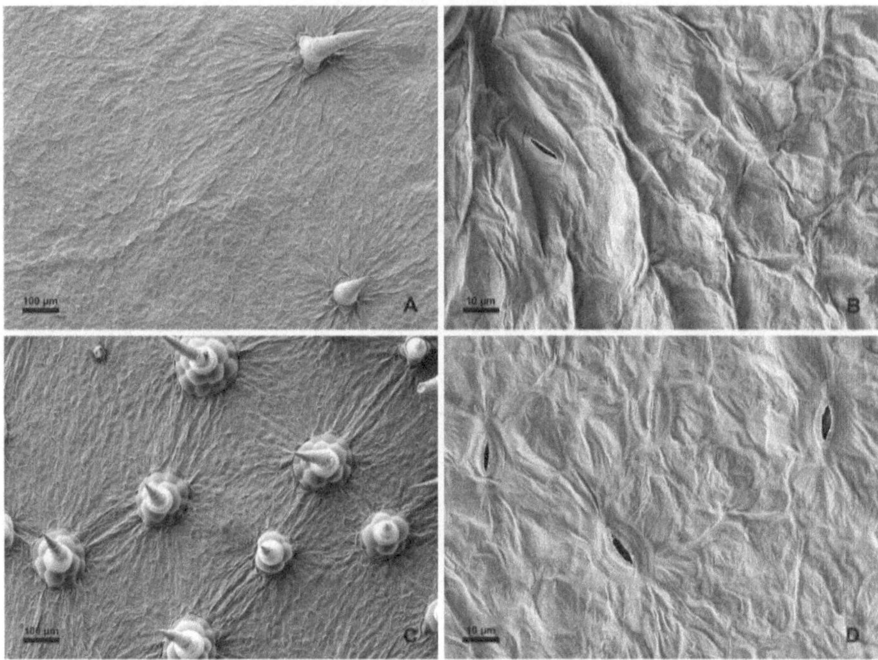

Fig. 8. The adaxial (A and B) and abaxial (C and D) surface of *Citrullus colocynthis* leaves processed by air-drying and observed under a scanning electron microscope.

The stomata of *P. dactylifera* were arranged in rows along the leaflet surface. The structure of the cuticular surface showed the absence of trichomes and the presence of cuticular wax crystalloids forming chimney-like structures surrounding the stomata (Fig. 9).

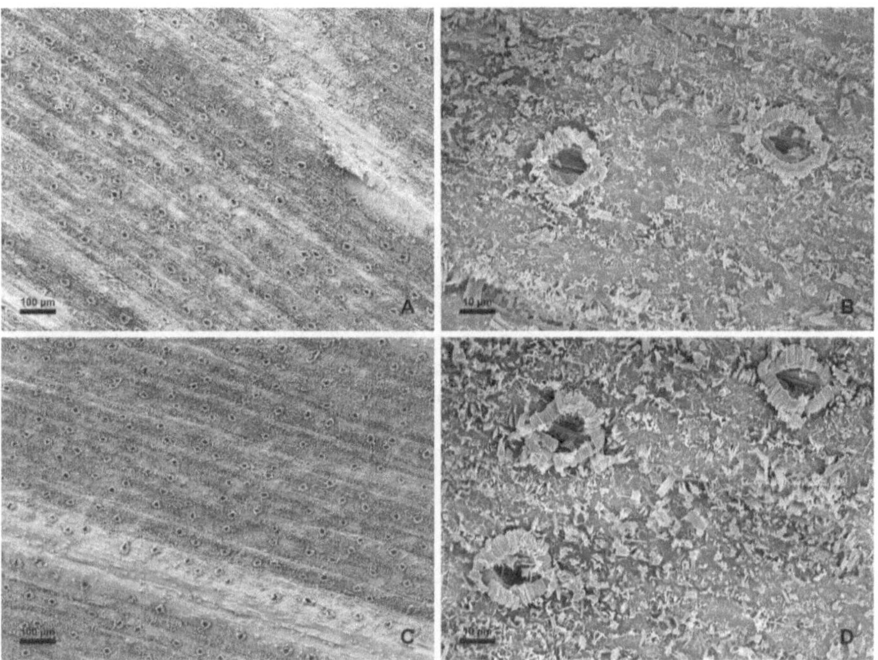

Fig. 9. The adaxial (A and B) and abaxial (C and D) surface of *Phoenix dactylifera* leaflets processed by air-drying and observed under a scanning electron microscope.

2.2.3 Leaf drying curves and minimum leaf water conductance

The actual leaf temperature at a given air temperature was considered in the conductance calculation because this value determines the driving force for water loss when the stomata are closed. Leaf temperatures of both *C. colocynthis* and *P. dactylifera* were slightly lower than the corresponding air temperatures at 25°C and 30°C. Raising air temperatures, the leaf-air temperature differences ($\Delta T_{leaf\text{-}air}$) of *C. colocynthis* continuously increased until reaching a value of -3.64°C at an air temperature of 50°C. The $\Delta T_{leaf\text{-}air}$ of *P. dactylifera* remained at around -1.41°C in all temperatures above 30°C (Table 4). Minimum cuticular transpiration (J_{min}) increased with air temperature. J_{min} rose from $1.27 \pm 0.37 \times 10^{-3}$ m s^{-1} to $13.16 \pm 2.80 \times 10^{-3}$ m s^{-1} in *C. colocynthis* and from $0.21 \pm 0.04 \times 10^{-3}$ m s^{-1} to $1.04 \pm 0.21 \times 10^{-3}$ m s^{-1} in *P. dactylifera* (Table 4).

Table 4. Leaf to air temperature difference and minimum transpiration (J_{min}) of *Citrullus colocynthis* leaves and *Phoenix dactylifera* leaflets obtained from drying curves depend on air temperature (T_{air}). Each value represents the mean ± standard deviation ($n \geq 8$).

T_{air} (°C)	$J_{min} \times 10^{-3}$ (m s^{-1})		$\Delta T_{leaf\text{-}air}$ (°C)	
	C. colocynthis	*P. dactylifera*	*C. colocynthis*	*P. dactylifera*
25	1.27 ± 0.37	0.21 ± 0.04	-0.47 ± 0.15	-0.33 ± 0.13
30	1.85 ± 0.29	0.25 ± 0.04	-0.50 ± 0.15	-0.56 ± 0.11
35	2.54 ± 0.54	0.40 ± 0.05	-0.88 ± 0.20	-1.04 ± 0.33
40	3.83 ± 0.99	0.45 ± 0.12	-1.02 ± 0.54	-1.30 ± 0.69
45	6.87 ± 2.38	0.65 ± 0.13	-2.34 ± 0.61	-1.83 ± 1.40
50	13.16 ± 2.80	1.04 ± 0.21	-3.64 ± 0.45	-1.47 ± 0.10

The cuticular water loss as a function of dehydration time from detached leaves of *C. colocynthis* and leaflets of *P. dactylifera* exposed to dry air were measured. Transpiration rates were calculated for each pair of consecutive data points along the leaf drying curve. For both plant species at 25°C, a high conductance characterised the initial phase of the leaf drying curves. Maintaining leaf dehydration, the conductance values declined until reaching a plateau being insensitive to a progressive decline of RWD. This constant low conductance represents the minimum conductance (g_{min}) at maximum stomatal closure. Leaf drying curves performed at 50°C showed the same pattern for *P. dactylifera* but not for *C. colocynthis*. The conductance of *C. colocynthis* at 50°C, after the initial phase, progressively decreased until a late stomatal closure when the RWD reaches values about 0.60 (Fig. 10). The transition point, which represents the RWD at which stomata were maximally closed (RWD$_{sc}$), was calculated from the intersection of the declining phase and the plateau phase of the leaf drying curves. The plant species showed different responses concerning the stomatal closure. In the temperature range from 25°C to 40°C, *C. colocynthis* maximally closed the stomata at RWDs between 0.10 and 0.20.

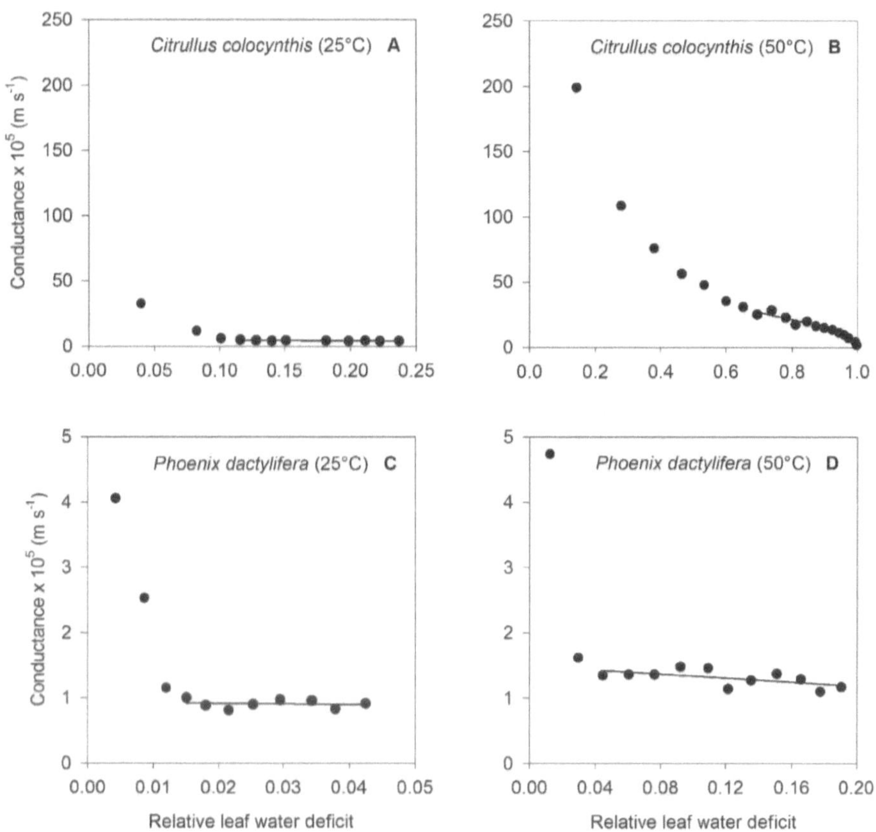

Fig. 10. Leaf conductance to water vapour as a function of the relative water deficit (RWD). The curves represent an individual measurement obtained from drying curves of *Citrullus colocynthis* leaves (A and B) and *Phoenix dactylifera* leaflets (C and D) at 25°C and 50°C. The RWD at maximum stomatal closure is marked by the transition between the declining phase and the plateau phase of leaf conductance. After a maximum stomatal closure, indicated by a linear curve, leaf conductance remained at a constant minimum and is defined as the minimum conductance.

At higher temperatures, maximum stomatal closure shifted to higher RWD up to 0.60 at an air temperature of 50°C. In comparison, *P. dactylifera* maximally closed the stomata at RWD ten times lower than in *C. colocynthis*, regardless of the air temperature increase (Table 5).

The minimum conductance of *C. colocynthis* at 25°C, $6.94 \pm 2.00 \times 10^{-5}$ m s^{-1}, was more than six-fold higher (t (14.6) = -10.8, $p < 0.001$) compared to the minimum conductance of *P. dactylifera*, $1.11 \pm 0.24 \times 10^{-5}$ m s^{-1}. The minimum conductance of *C. colocynthis* increased with temperature by a factor of 3.17 between 25°C and 50°C from $6.94 \pm 2.00 \times 10^{-5}$ m s^{-1} to $21.98 \pm 5.19 \times 10^{-5}$ m s^{-1}. In comparison, the minimum conductance of *P. dactylifera* increased only by a factor of 1.17 from $1.11 \pm 0.24 \times 10^{-5}$ m s^{-1} to $1.30 \pm 0.26 \times 10^{-5}$ m s^{-1} (Table 5).

Table 5. Minimum conductance (g_{min}) and relative water deficit at maximum stomatal closure (RWD$_{SC}$) of *Citrullus colocynthis* leaves and *Phoenix dactylifera* leaflets obtained from leaf drying curves depend on air temperature (T$_{air}$). Each value represents the mean ± standard deviation ($n \geq 8$).

T$_{air}$ (°C)	$g_{min} \times 10^{-5}$ (m s^{-1})		RWD$_{SC}$	
	C. colocynthis	*P. dactylifera*	*C. colocynthis*	*P. dactylifera*
25	6.94 ± 2.00	1.11 ± 0.24	0.14 ± 0.05	0.014 ± 0.004
30	7.83 ± 1.23	0.92 ± 0.16	0.15 ± 0.06	0.010 ± 0.006
35	7.74 ± 1.46	1.00 ± 0.12	0.16 ± 0.03	0.014 ± 0.001
40	8.79 ± 2.41	1.19 ± 0.47	0.20 ± 0.08	0.011 ± 0.006
45	13.70 ± 5.33	1.37 ± 0.32	0.30 ± 0.08	0.015 ± 0.003
50	21.98 ± 5.19	1.30 ± 0.26	0.60 ± 0.04	0.016 ± 0.005

The minimum conductance of *C. colocynthis* slightly increased between 25°C to 40°C followed by a steep increase (x^2 (5) = 35.6, $p < 0.001$) at temperatures above 40°C. Nevertheless, the minimum conductance of *P. dactylifera* sparingly increased from 25°C to 50°C (x^2 (5) = 21.1, $p = 0.01$) without an abrupt change in slope (Fig. 11).

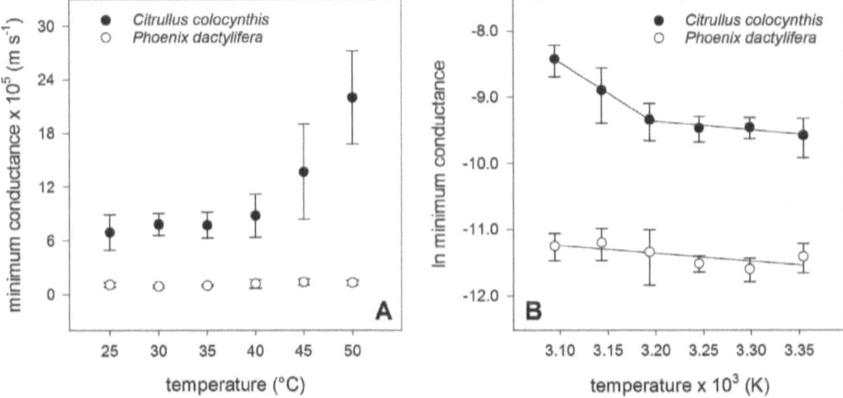

Fig. 11. Minimum conductance (g_{min}, A) and Arrhenius plots of the g_{min} (B) of *Citrullus colocynthis* leaves and *Phoenix dactylifera* leaflets obtained from drying curves as a function of air temperature. The rising of temperature from 25°C to 50°C results in a minimum conductance increase factor of 3.2 for *C. colocynthis* and 1.2 for *P. dactylifera*. Each value represents the mean ± standard deviation ($n \geq 8$).

2.2.4 Chemical composition of cuticular waxes

The cuticular waxes of *C. colocynthis* leaves and *P. dactylifera* leaflets were analysed qualitatively and quantitatively to investigate differences between both plant species and detecting potential relationships between cuticular transpiration and the cuticular wax composition. The leaf cuticular wax coverage was 4.19 ± 0.44 µg cm^{-2} for *C. colocynthis* and 29.35 ± 4.24 µg cm^{-2} for *P. dactylifera* (Table 6). The cuticular waxes of both plant species were mainly composed of very-long-chain aliphatic compounds (Fig. 12).

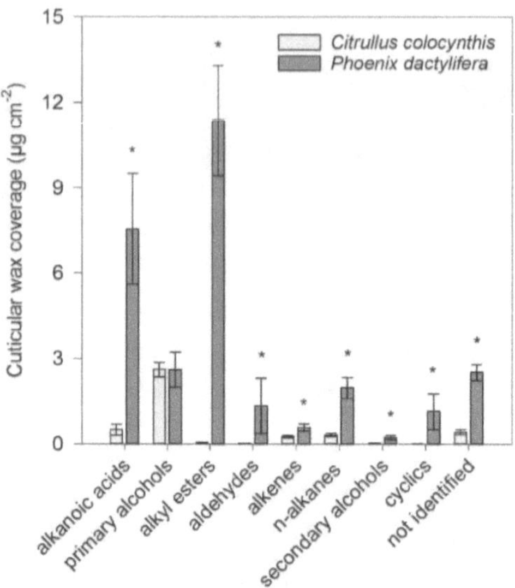

Fig. 12. Cuticular wax coverage of *Citrullus colocynthis* leaves and *Phoenix dactylifera* leaflets arranged by compound class. Each value represents the mean ± standard deviation ($n \geq 3$). Asterisk indicates significant difference of the wax coverage between the two species (p < 0.05).

Very-long-chain aliphatics with carbon chain lengths ranging from C_{20} to C_{46} amounted to 90% of total cuticular waxes in *C. colocynthis,* and the main constituents were the primary alcohols octacosanol (C_{28}; 13% of total cuticular waxes), triacontanol (C_{30}; 26%) and dotriacontanol (C_{32}; 13%). In *P. dactylifera,* the very-long-chain aliphatics with carbon chain lengths ranging from C_{20} to C_{62} amount to 88% of total cuticular waxes with dotriacontanoic acid (C_{32}; 15% of total cuticular waxes) and alkyl esters with carbon chain length C_{48} (10%), C_{56} (6%), and C_{58} (6%) as the dominant constituents. Only a minor fraction of cuticular waxes contained cyclic compounds in both *C. colocynthis* (< 1% of total cuticular waxes) and *P. dactylifera* (4%).

Table 6. The cuticular wax coverage and components of *Citrullus colocynthis* leaves and *Phoenix dactylifera* leaflets. Each value represents the mean ± standard deviation ($n \geq 3$).

Compound class	Carbon chain length	Wax coverage ($\mu g\ cm^{-2}$)	
		C. colocynthis	*P. dactylifera*
alkanoic acids	20	0.003 ± 0.001	-
	21	0.001 ± 0.001	0.005 ± 0.001
	22	0.014 ± 0.020	0.023 ± 0.003
	23	-	0.012 ± 0.003
	24	0.014 ± 0.002	0.088 ± 0.043
	25	0.005 ± 0.003	0.019 ± 0.008
	26	0.022 ± 0.004	0.104 ± 0.025
	27	0.009 ± 0.004	0.041 ± 0.020
	28	0.055 ± 0.016	0.273 ± 0.054
	29	0.019 ± 0.004	0.096 ± 0.025
	30	0.176 ± 0.073	0.581 ± 0.134
	31	0.019 ± 0.007	0.357 ± 0.069
	32	0.128 ± 0.068	4.519 ± 1.735
	33	0.006 ± 0.001	0.173 ± 0.043
	34	0.018 ± 0.010	1.264 ± 0.230
	36	0.010 ± 0.001	-
primary alcohols	22	0.002 ± 0.001	0.002 ± 0.000
	24	0.007 ± 0.002	0.014 ± 0.003
	25	0.002 ± 0.001	0.007 ± 0.003
	26	0.047 ± 0.012	0.072 ± 0.015
	27	0.018 ± 0.003	0.014 ± 0.005
	28	0.563 ± 0.063	0.260 ± 0.114
	29	0.081 ± 0.007	0.013 ± 0.010
	30	1.101 ± 0.122	0.162 ± 0.035
	31	0.055 ± 0.005	0.169 ± 0.022
	32	0.543 ± 0.048	1.223 ± 0.401
	33	0.024 ± 0.002	0.103 ± 0.007
	34	0.123 ± 0.014	0.574 ± 0.188
	35	0.007 ± 0.001	-
	36	0.046 ± 0.007	-
alkyl esters	40	-	0.029 ± 0.008
	42	0.012 ± 0.002	0.080 ± 0.010
	43	-	0.016 ± 0.006
	44	0.010 ± 0.002	0.116 ± 0.033
	45	-	0.028 ± 0.008
	46	0.022 ± 0.004	0.376 ± 0.155
	47	-	0.075 ± 0.017

Compound class	Carbon chain length	Wax coverage ($\mu g\ cm^{-2}$)	
		C. colocynthis	*P. dactylifera*
	48	-	2.906 ± 0.262
	49	-	0.124 ± 0.012
	50	-	1.414 ± 0.394
	51	-	0.060 ± 0.022
	52	-	0.440 ± 0.083
	53	-	0.083 ± 0.022
	54	-	0.769 ± 0.028
	55	-	0.128 ± 0.040
	56	-	1.742 ± 0.346
	57	-	0.173 ± 0.052
	58	-	1.616 ± 0.510
	59	-	0.096 ± 0.052
	60	-	0.870 ± 0.323
	61	-	0.041 ± 0.032
	62	-	0.192 ± 0.075
	30	0.019 ± 0.003	-
aldehydes	31	-	0.052 ± 0.031
	32	-	1.111 ± 0.861
	34	-	0.184 ± 0.080
	26	0.008 ± 0.002	0.020 ± 0.005
	27	-	0.023 ± 0.005
	28	0.092 ± 0.015	0.050 ± 0.018
alkenes	29	-	0.094 ± 0.018
	30	0.088 ± 0.010	0.266 ± 0.069
	32	0.047 ± 0.004	0.126 ± 0.051
	34	0.024 ± 0.002	-
	25	0.011 ± 0.006	-
	26	-	0.011 ± 0.002
	27	0.036 ± 0.021	0.023 ± 0.007
	28	0.007 ± 0.002	0.031 ± 0.010
n-alkanes	29	0.095 ± 0.021	0.063 ± 0.031
	30	0.009 ± 0.001	0.103 ± 0.030
	31	0.110 ± 0.012	1.118 ± 0.308
	32	-	0.074 ± 0.031
	33	0.040 ± 0.004	0.557 ± 0.047
	27	0.003 ± 0.001	-
secondary alcohols	31	0.014 ± 0.009	0.152 ± 0.047
	33	0.009 ± 0.006	0.065 ± 0.024
total very-long-chain aliphatics		*3.771 ± 0.392*	*25.664 ± 3.654*
alpha-amyrin			0.439 ± 0.322
beta-amyrin		-	0.116 ± 0.092

Compound class	Carbon chain length	Wax coverage ($\mu g\ cm^{-2}$)	
		C. colocynthis	*P. dactylifera*
lupeol		-	0.253 ± 0.054
lupenon		-	0.084 ± 0.060
triterpenoid I		-	0.181 ± 0.048
triterpenoid II		-	0.071 ± 0.041
cholesterol		0.002 ± 0.001	-
total cyclic aliphatics		*0.002 ± 0.001*	*1.145 ± 0.624*
beta-tocopherol		0.005 ± 0.003	-
total cyclic aromatics		*0.005 ± 0.003*	-
total cyclics		*0.008 ± 0.003*	*1.145 ± 0.624*
not identified		0.415 ± 0.081	2.530 ± 0.226
total wax		*4.193 ± 0.436*	*29.338 ± 4.235*

2.2.5 Chemical composition of the cutin matrix

The amount of the cutin monomers of *C. colocynthis* leaves was 7.75 ± 1.28 $\mu g\ cm^{-2}$ and of *P. dactylifera* leaflets 57.44 ± 4.18 $\mu g\ cm^{-2}$ (Table 7). For both plant species, the cutin matrix composed of about 88% aliphatic cutin monomers and about 12% phenolic cutin monomers. For the leaf cutin of *C. colocynthis*, the carbon chain lengths of aliphatic compounds ranged from C_{16} to C_{32} but up to C_{34} for *P. dactylifera* leaflets. The ratio of the cutin acids with a carbon chain length of C_{16} and C_{18} averaged 6:1 for *C. colocynthis* leaves, and 1:11 for *P. dactylifera* leaflets. The predominant cutin acid was 9/10,ω-dihydroxy hexadecanoic acid (51% of total cutin monomers) in *C. colocynthis* leaves and ω-hydroxy 9,10-epoxy octadecanoic acid (60%) in *P. dactylifera* leaflets. Ω-hydroxy 9,10-epoxy octadecanoic acid was absent in the *C. colocynthis* leaf cutin. The amount of phenolic cutin acids based mainly on an accumulation of *trans*-4-hydroxycinnamic acid (*para*-coumaric acid) with 7% of total cutin monomers in *C. colocynthis* and *P. dactylifera*, respectively.

Table 7. The cutin monomeric coverage of *Citrullus colocynthis* leaves and *Phoenix dactylifera* leaflets. Each value represents the mean ± standard deviation ($n \geq 5$).

Compound	Carbon chain length	Cutin coverage (µg cm^{-2})					
		C. colocynthis			*P. dactylifera*		
alkanoic acid	16	0.40	±	0.19	0.55	±	0.05
alkanoic acid	17	0.03	±	0.01	-		
alkatrienoic acid (9,12,15)	18:3	0.16	±	0.02	-		
alkadienoic acid (9,12)	18:2	0.10	±	0.02	-		
alkenoic acid (9)	18:1	0.09	±	0.02	-		
alkanoic acid	18	0.22	±	0.06	0.29	±	0.15
alkanoic acid	20	0.03	±	0.01	0.01	±	0.00
alkanoic acid	22	0.05	±	0.01	0.07	±	0.01
alkanoic acid	24	0.04	±	0.01	0.60	±	0.05
alkanoic acid	26	-			0.61	±	0.03
alkanoic acid	28	-			0.22	±	0.02
alkanoic acid	30	0.03	±	0.01	0.40	±	0.05
alkanoic acid	32	0.03	±	0.01	0.12	±	0.02
alkanoic acid	34	-			0.12	±	0.03
primary alcohol	16	0.04	±	0.02	0.07	±	0.03
primary alcohol	18	0.05	±	0.01	0.02	±	0.01
primary alcohol	20	0.06	±	0.01	0.03	±	0.01
primary alcohol	22	0.06	±	0.00	-		
primary alcohol	24	0.03	±	0.01	-		
primary alcohol	26	0.01	±	0.01	0.03	±	0.01
primary alcohol	28	0.06	±	0.02	0.03	±	0.01
primary alcohol	30	0.05	±	0.01	0.13	±	0.04
primary alcohol	32	-			0.26	±	0.02
primary alcohol	34	-			0.03	±	0.00
9/10-hydroxy alkane-1,16-dioic acid	16	0.40	±	0.12	-		
16-hydroxy alkenoic acid (9)	16:1	0.04	±	0.01	-		
16-hydroxy alkanoic acid	16	0.28	±	0.03	0.08	±	0.01
18-hydroxy alkadienoic acid (12,15)	18:2	-			0.66	±	0.04
18-hydroxy alkadienoic acid (9,12)	18:2	-			1.14	±	0.08
18-hydroxy alkenoic acid (9)	18:1	-			0.70	±	0.07
18-hydroxy alkanoic acid	18	0.06	±	0.01	-		
9/10,16-dihydroxy alkanoic acid	16	3.97	±	0.82	3.11	±	0.23
9/10,18-dihydroxy alkanoic acid	18	0.23	±	0.05	0.43	±	0.05
16-hydroxy 9/10-oxo alkanoic acid	16	-			0.03	±	0.00
18-hydroxy 9/10-oxo alkanoic acid	18	-			0.25	±	0.02
18-hydroxy 9,10-epoxy alkanoic acid	18	-			34.28	±	2.57
9,10,18-trihydroxy alkanoic acid	18	-			4.87	±	0.92
2-hydroxy alkanoic acid	22	0.06	±	0.00	0.19	±	0.02
2-hydroxy alkanoic acid	23	0.03	±	0.01	0.03	±	0.01

Compound	Carbon chain length	Cutin coverage ($\mu g\ cm^{-2}$)					
		C. colocynthis			*P. dactylifera*		
2-hydroxy alkanoic acid	24	0.22	±	0.02	0.31	±	0.10
2-hydroxy alkanoic acid	25	0.04	±	0.01	0.06	±	0.00
2-hydroxy alkanoic acid	26	0.07	±	0.01	-		
cis-4-hydroxy cinnamic acid		0.04	±	0.01	0.39	±	0.07
trans-4-hydroxy cinnamic acid		0.52	±	0.09	3.85	±	0.10
3,4-dihydroxy cinnamic acid		0.22	±	0.11	0.06	±	0.06
4-hydroxy 3-methoxy cinnamic acid		0.01	±	0.00	0.01	±	
3,5-dimethoxy 4-hydroxy cinnamic acid		-			0.28	±	0.09
4-hydroxy benzoic acid		-			1.55	±	0.07
3,4-dihydroxy benzoic acid		-			0.57	±	0.04
4-hydroxy 3-methoxy benzoic acid		0.05	±	0.02	0.18	±	0.02
3-methoxy mandelic acid		-			0.41	±	0.02
4-hydroxy 3-methoxy mandelic acid		-			0.45	±	0.18
total cutin		*7.75*	*±*	*1.28*	*57.44*	*±*	*4.18*

2.3 Discussion

The life strategies adopted by *Citrullus colocynthis* and *Phoenix dactylifera* reflect their growth form and functional diversification driven by spatial variability in abiotic filters. *C. colocynthis* is a broad-summer-green vine, which spreads its trailing stems along the soil surface. Its leaves stand only a few centimetres above the ground (Lange, 1959). Stem growth begins from buds on the taproots of established plants, and after reproduction, i.e. fruit production, the aboveground parts die off (Cunningham *et al.*, 2011). *P. dactylifera*, on the other hand, is a long-lived evergreen woody plant and can grow up to 30 m height. The straight leaves with obliquely vertical orientation form the tree crown at the end of the trunk (Barrow, 1998).

In an ecological context, plant leaf size plays a central role in leaf energy and water balance, while LMA reflects the trade-off between carbon gain and longevity (Díaz *et al.*, 2016). Also, the leaf water content (LWC) roughly indicates leaf density (Garnier and Laurent 1994). The evergreen *P. dactylifera* had, as expected, an LMA three-fold higher than that of *C. colocynthis*. LMA calculation did not include petioles because leaflets were used for performing the leaf drying curves. The high costs for producing conservative evergreen leaves have to be compensated by a higher survival rate in the face of abiotic and biotic hazards (Coley *et al.*, 1985; Reich *et al.*, 1997; Wright *et al.*, 2004). Hence, the high LMA and low LWC may confer higher stress resistance to *P. dactylifera* when compared with *C. colocynthis*.

Desert soil surfaces can reach extreme temperatures due to the intense solar radiation. Lange (1959) reported soil temperatures above 70°C in the Sahara desert. Living near the desert soil can be very problematic for plants, especially for seedlings and trailing plants; thereby eco-morpho-physiological adaptations might be required to cope with the heat stress. Lange (1959) measured temperatures of *C. colocynthis* leaves, *P. dactylifera* leaflets and, corresponding air temperatures under extreme conditions in the Sahara desert. The leaf temperature of *C. colocynthis* in the morning was about 10°C cooler than that of air temperature. It maintained this low leaf temperature despite the intense irradiation throughout the day, and the air to leaf temperature difference rose up to 15°C in the early afternoon. The temperature of a cut leaf, after a short time, overstepped the air temperature and even reached temperatures higher than 60°C. The author associated this strong cooling effect with high transpiration and consequently a high water consumption of *C. colocynthis*.

In our leaf drying experiments we measured the cuticular transpiration of *P. dactylifera* and *C. colocynthis* under strictly controlled conditions. For *C. colocynthis* we found a minimum leaf conductance (g_{min}) of $6.94 \pm 2.00 \times 10^{-5}$ m s^{-1}, which fits appropriately with permeability values reported by Schuster *et al.* (2017) for three broad-summer-green vines. Also, it lies within the upper range of the cuticular permeability measured for dozens of plant species (10^{-7} to 10^{-4} m s^{-1}; Riederer Schreiber, 2001). Thus, the cuticular barrier to water loss of *C. colocynthis* is not particularly efficient. Furthermore, the relative water deficit at maximum stomatal closure (RWD_{sc}) was at 0.15 and drastically increased above 40°C to about 0.60 at 50°C (Table 3). In the field, the cooling effect of the stomatal transpiration keeps the leaf temperatures below the thermal tolerance of this plant species (46°C), avoiding death by overheating (Lange, 1959). This strategy distinguishes *C. colocynthis* as a typical water-spender plant highly adapted to heat stress (Lange, 1959; Althawadi and Grace, 1986; Beerling *et al.*, 2001; Schulze *et al.*, 2005). While decreasing the leaf water potential, it maintains the water flow from the soil to compensate the water loss by stomatal transpiration, which is necessary for leaf cooling.

In comparison, Lange (1959) did not find a cooling effect for *P. dactylifera*. The leaflet temperature was up to 11°C above that of the air temperature, and the transpiration rates were almost unnoticeable. Furthermore, the leaflet temperature did not abruptly increase after being cut off from the plant. Although the leaflet had overheated, the leaflet temperature did not increase over the thermal limit of *P. dactylifera* (59°C; Lange, 1959). The g_{min} of *P. dactylifera* at 25°C, $1.11 \pm 0.24 \times 10^{-5}$ m s^{-1}, was about six-fold lower than that of *C. colocynthis*. It indicates that the cuticular transpiration barrier of *P. dactylifera* is more efficient to avoid water loss. The relative RWD_{sc} in *P. dactylifera* is ten times lower than for *C. colocynthis* and stays constant low over the whole temperature range measured (Table 3). Müller *et al.* (2017) demonstrated that *P. dactylifera* has strong stomatal control. The ability to close the stomata at small changes in the RWD at all experienced temperatures, preventing leaf damage and the lower minimum conductance minimise the water loss, specifying *P. dactylifera* as a typical water-saver plant. Maintaining the leaves in a vertical position far away from the soil surface and easily exposed to the wind may represent a morphological solution for optimising heat dissipation by convection in *P. dactylifera*, whereas the transpiration-cooling effect of *C. colocynthis* leaves may be an eco-physiological specialisation to deal with high temperatures near to the desert ground.

The cuticular water permeability is of paramount importance for maintaining a favourable water status of plants, especially in desert plants. This importance is in sharp contrast to the knowledge on the structure, chemical composition, and physical features of cuticles from desert plants. To date, there is only a single study concerning the cuticular water permeability of a desert plant and its dependence on chemical, physical, and environmental factors. Schuster *et al.* (2016) showed that *Rhazya stricta*, a desert shrub, does not possess a specially effective cuticular barrier to water loss at moderate temperatures but g_{min} did not drastically increase above a specific temperature, as is also true for *P. dactylifera*. The cuticular wax of *Rhazya stricta* is mainly composed of thermo-stable triterpenoids. Also, a high triterpenoid-to-cutin ratio was obtained. Therefore, Schuster *et al.* (2016) proposed, *"that the triterpenoids deposited within the cutin matrix restrict the thermal expansion of the polymer and, thus, prevent thermal damage to the highly ordered aliphatic wax barrier even at high temperatures"*.

The knowledge of potential functions of very-long-chain aliphatic compounds has expanded over the last years. Experiments with *Solanum lycopersicum* fruits (Vogg *et al.*, 2004; Leide *et al.*, 2007, 2011), *Arabidopsis thaliana* leaves (Buschhaus and Jetter, 2012) and leaves of eight evergreen woody plant species (Jetter and Riederer, 2016), suggest that the contribution of very-long-chain aliphatics to the efficiency of the transpiration barrier is crucial. These findings corroborate the model for the molecular structure of cuticular waxes proposed by Riederer and Schreiber (1995). The authors suggested that cuticular waxes are multiphase systems composed of amorphous zones within highly ordered crystalline zones. According to this model, the very-long-chain aliphatic molecules correspond to the crystalline domains of the cuticular waxes, while the amorphous zones should incorporate the chain ends and cyclic components. This model predicts that the very-long-chain aliphatic fraction of the cuticular waxes mainly represents the cuticular transpirational barrier in plants. The cuticular waxes of both *P. dactylifera* and *C. colocynthis* fit within this model. Although *P. dactylifera* and *C. colocynthis* qualitatively present similar aliphatic compound classes, the quantitative contribution of each class for total cuticular wax load and the carbon chain length of homologous compounds considerably vary between these plant species. The cuticular waxes of *P. dactylifera* contain higher amounts of all identified compound classes, except for the equivalence of primary alcohols, and the total very-long-chain aliphatic wax load was seven-fold higher than in *C. colocynthis*.

Although the carbon chain lengths of the very-long-chain aliphatic compounds range from C_{20} to C_{62} in *P. dactylifera* and from C_{20} to C_{42} in *C. colocynthis*, the weighted median carbon chain length (MCL) of the aliphatic cuticular wax components only differ by about three carbon atoms (31.8 and 29.4 in *P. dactylifera* and *C. colocynthis*, respectively). Nevertheless, the patterns of chain length distribution of very-long-chain aliphatic wax components drastically change between the two species (Fig. 13).

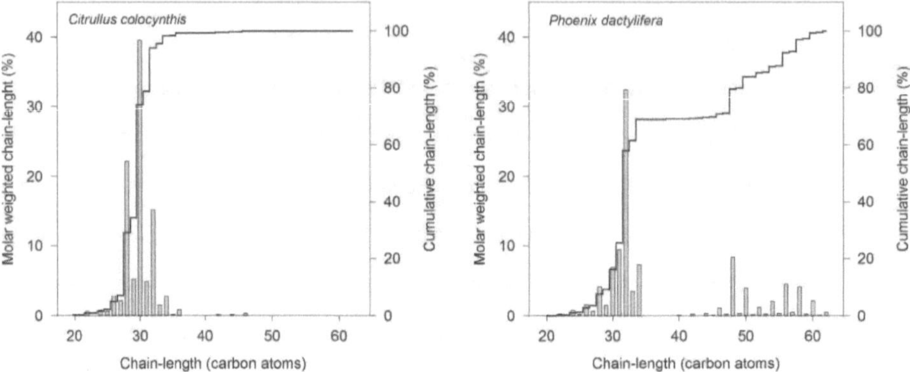

Fig. 13. Individual and cumulative chain-length distributions of cuticular waxes of *Citrullus colocynthis* leaves (left) and *Phoenix dactylifera* leaflets (right).

According to the chain length, the very-long-chain aliphatics of both species can be divided into components shorter than 40 carbon atoms and those equal or longer. Within the components shorter than C_{40} are alkanes, alkenes, aldehydes, alkanoic acids and, primary and secondary alcohols. Moreover, the alkyl esters exclusively make up the very-long-chain aliphatic fraction longer than C_{40}. The weighted aliphatic components shorter than C_{40} comprise 69% of the total very-long-chain aliphatic fraction in *P. dactylifera*, and those equal or longer than C_{40} contribute with 31%. In contrast, only a minor fraction (1%) of the very-long-chain aliphatic components were longer than C_{40} in *C. colocynthis*. This difference becomes even more apparent when the MCL of very-long-chain aliphatics above C_{40} are compared, i.e. the MCL is over seven carbon atoms longer in *P. dactylifera* (51.3) than that of *C. colocynthis* (43.9). Hence, the high amount and the outstanding MCL of the very-long-chain aliphatic compounds in *P. dactylifera* might increase the size and alter the geometry of cuticular wax crystalline domains, thereby limiting more efficiently the water flow through its cuticle.

The melting point of homologous aliphatic compounds rises with increasing carbon numbers, and one may expect that the high proportion of very-long-chain alkyl esters contributes to better thermal stability in the *P. dactylifera* cuticle. Iyengar and Schlenk (1969) determined the melting point of saturated wax esters up to C_{40} and showed that the melting point of arachidyl arachidate (C_{40}) is not unusually high (69 °C). For instance, the melting points of dotriacontanoic acid (C_{32}) and triacontanol (C_{30}) are 96 and 87°C, respectively. However, considering the wax structure proposed by Reynhardt and Riederer (1994), the very-long-chain alkyl esters play an essential role in the physical properties of plant waxes. The authors suggest that the very-long-chain alkyl esters might be oriented parallel to shorter chains, linking adjacent crystalline domains within the mobile amorphous zone. Thus, despite the relatively low melting points reported to single homologous of alkyl esters, the role played in connecting crystalline zones might increase the melting point of the total wax considerably. Thereby, conferring high thermal stability to *P. dactylifera* cuticle.

It was also proposed that the cutin polymer plays a significant role in establishing an effective physical barrier by providing a framework, in which the cuticular waxes are arranged (Kosma and Jenks, 2007). The cutin matrix of both plant species mainly consisted of ester-linked C_{16} and C_{18} hydroxy alkanoic acids predominantly 9/10,ω-dihydroxy hexadecanoic acid in *C. colocynthis* leaves and ω-hydroxy 9,10-epoxy octadecanoic acid in *P. dactylifera* leaflets. A detailed comparison of the cutin monomers showed that the ratio of C_{16} and C_{18} cutin acids was distinctly different. The cutin composition of *C. colocynthis* leaves represents a C_{16} type, and *P. dactylifera* leaflets a C_{18} type of cutin (Holloway, 1982). In addition to this differential distribution of carbon chain lengths, the cutin amount was seven-fold lower in *C. colocynthis* leaves compared to *P. dactylifera* leaflets. Compositional dissimilarities of leaf cutin monomers might indicate a unique cross-linking within the polymeric cutin structure of both plant species resulting from differential carbon chain length distribution, the degree of unsaturation and degree of epoxidation of cutin acid monomers.

In conclusion, our findings showed that the cuticle of the water-saver *P. dactylifera* when compared with the water-spender *C. colocynthis*: (i) is more efficient to avoid cuticular transpirational water loss, (ii) is more stable under heat stress, and (iii) features differentiated cuticular wax coverage and cutin deposition. The cuticular waxes play an essential role in the efficiency and stability of the transpirational barrier to water loss. We suggest that the evergreen leaves of *P. dactylifera* have a more efficient cuticular transpiration barrier maintained by very-long-chain aliphatic wax compounds with remarkable high MCL embedded within a C_{18} type cutin matrix. We also provide substantial evidence of adaptation at the cuticular level related to the different life strategy of the two desert plants. Thus, future studies addressing the biology of the plant cuticle should address the plant life strategy in a broader ecological context instead of considering merely the plant habitat.

Chapter 3

Coevolution of photosynthetic thermal tolerance and cuticular transpiration barrier in non-succulent xerophytes

Abstract: Drought and heat represent two of the most prominent land plant stressors. They often go together in a wide range of ecosystems, and the successful plant colonisation of the hot arid regions across the globe might have required a synergic adaptation to drought and heat stresses. Non-succulent xerophytes acquired through evolution many morphological and physiological modifications, including leaf photosynthetic thermal tolerance and traits related to water balance, which allowed them to deal with drought and heat. This work aims to evaluate whether plant adaptations to cope with drought (at cuticular level) and heat (at photosynthetic level) coevolved under three hot arid biomes (Saudi desert, Portuguese Mediterranean, and Brazilian tropical dry forest). We hypothesised that the cuticular transpirational barrier of the non-succulent xerophytes close relates to their leaf thermal tolerance. We tested this Hypothesis using the minimum leaf water conductance (g_{min}) as a proxy for cuticular water permeability and the critical temperature (T_{crit}) of the photosystem II as an indicator or leaf thermal tolerance. The g_{min} of the xerophilous plants ranged from $0.46 \pm 0.24 \times 10^{-5}$ m s^{-1} for *S. japicanga* to $6.94 \pm 2.00 \times 10^{-5}$ m s^{-1} for *C. colocynthis*. Although g_{min} varied at species-specific level by more than 14-fold, there was no difference among habitats. Similarly, T_{crit} differed among species by more than 6°C, but no difference between habitats was identified. Also, we provide substantial evidence that non-succulent xerophytes with more efficient cuticular transpiration barrier possess higher leaf thermal tolerance, which indicates potential coevolution of these features in hot arid biomes.

Keywords: Hot desert, Mediterranean woodland, dry forest, minimum leaf conductance, leaf heat tolerance, arid climate.3.1 Introduction

3.1 Introduction

Climate change might increase the global temperatures in circa 2°C (IPCC, 2014), and warming consequences to deserts might be more comprehensive than making hot-deserts hotter (Ward, 2016). Studies indicate that the size of deserts might increase with the predicted climate changes (Volder *et al.,* 2010; Engelbrecht and Engelbrecht, 2016; Ward, 2016). Also, Perkins *et al.* (2012) estimated that the frequency and intensity of heat waves and summer droughts would increase across the globe. This scenario is also corroborated by Trenberth *et al.* (2014), which predicted a quicker establishment and prolonged droughts due to the extra heat from global warming. Thus, the challenge of living in arid regions would be intensified to plants due to the expected hotter and drier environment. Drought and heat are two of the most significant plant stressors since the first plants have colonised the land. Moreover, both stressors often co-occur in a wide range of ecosystems, and the successful plant colonisation of the hot arid regions over the world might have required a synergic adaptation to drought and heat stresses.

Flowering plants possess high adaptability and tolerance to a broad gradient of temperature and precipitation (Knight and Ackerly, 2003). These plants are present within the entire continuum of habitats, from those with few weeks of frost-free weather to hot deserts (Knight and Ackerly, 2003). The evolution to occupy dry regions of the world with concomitantly high temperatures required many morphological and physiological features, which include leaf photosynthetic thermal tolerance (P_{TT}) and traits related to water balance. Similar to many other physio and biochemical process, photosynthesis promptly responds to temperature (Knight and Ackerly, 2003), and the evidence available indicates that the photosystem II (PSII) is one of the most thermally labile constituents of photosynthesis (Weis and Berry, 1988; Havaux, 1993). Hence, we used the excitation capacity of PSII as an index of P_{TT}.

High temperatures often go along with drought in arid climate regions. Drought induces plant stomata closure, thereby minimising the water loss by stomatal transpiration. However, even with stomata maximally closed, there is still an remained uncontrolled water loss from the plant mesophyll to the dry atmosphere across the cuticle. The plant cuticle cover the outer epidermal cell wall of leaves and fruits and constitutes the interface between plant organs and their surrounding atmosphere (Riederer and Schreiber, 2001). Within limited water availability environments, the cuticular transpiration in relation to the plant water reservoirs plays a vital role in the plant fitness and survival (Schuster *et al.*, 2017) . In some plant species, the cuticle can be isolated enzymatically and used for permeability measurements (Schönherr and Riederer, 1986). However, this approach is only appropriate for astomatous and trichome-free cuticular membranes, thereby limiting the number of suitable plant species. Leaf drying curve is an alternative method to circumvent this problem. In this method, dry air and darkness are used to induce stomata closure of intact detached leaves, and we can readily determine the minimum leaf water conductance (g_{min}) gravimetrically from leaf drying curves. The cuticular permeability corresponds to the lowest water loss a whole leaf reaches when its stomata are maximally closed due to desiccation stress (Körner, 1995). Therefore, we used g_{min} as a proxy of cuticular transpiration barrier.

The analysis of xerophilous flowering plants exposed to extreme climate conditions in their natural environment appears to be a promising approach to trace potential clues of the coevolution of physiological traits to cope with drought and heat. Moreover, it can also lead to the characterisation of adaptative features that can potentially reduce the sensitivity of plants to climate change (by preventing or diminishing injurious effects of increased temperature and drought). This work aims to evaluate whether plant adaptations to cope with water (at cuticular level) and heat (at photosynthetic level) stresses coevolved under three hot arid biomes. We hypothesised that the cuticular transpirational barrier of the non-succulent xerophytes is closely related to their leaf photosynthetic thermal tolerance. From this hypothesis, we predicted the cuticular water permeability to be lower in high heat tolerant plants, i. e. the higher the leaf thermal tolerance, the better the efficiency of the cuticular transpiration barrier. We tested this Hypothesis using the minimum leaf conductance (g_{min}) as a proxy for cuticular water permeability and the critical temperature (T_{crit}) of the photosystem II as an indicator or leaf thermal tolerance.

3.2 Results

3.2.1 Minimum leaf conductance

Minimum leaf conductances (g_{min}) with maximal stomatal closure were investigated at 25°C from drying curves of fully developed leaves ($n \geq 9$) and used as an indicator of the efficiency of the cuticular transpiration barrier. The first stage of leaf drying curves were characterised by high conductances that decreased with leaf dehydration until reaching a plateau of constant conductance values when stomata maximally close. The g_{min} of the xerophilous plants ranged from $0.46 \pm 0.24 \times 10^{-5}$ m s^{-1} (mean $\pm$ standard deviation) for *S. japicanga* (a dry forest plant) to $6.94 \pm 2.00 \times 10^{-5}$ m s^{-1} *C. colocynthis* (a desert plant). For more details, see Table 8. Although g_{min} varied at species-specific level by more than 14-fold, there was no difference ($F (2) = 1.06$, $p = 0.379$) in g_{min} among habitats (Fig. 14).

Table 8. Minimum leaf conductances (g_{min}) of 14 flowering plant species from three hot and dry biomes. Each value represents the mean $\pm$ standard deviation ($n \geq 9$ biological replicates).

Plant species	Original habitat	$g_{min} \cdot 10^5$ (m s^{-1})
Byrsonima gardneriana	Tropical dry forest	1.97 ± 0.47
Calotropis procera	Hot desert	2.97 ± 0.66
Ceratonia siliqua	Mediterranean woodland	1.49 ± 0.26
Chamaerops humilis	Mediterranean woodland	1.46 ± 0.16
Citrullus colocynthis	Hot desert	6.94 ± 2.00
Cynophalla flexuosa	Tropical dry forest	2.34 ± 0.32
Olea europaea	Mediterranean woodland	2.63 ± 0.61
Phillyrea angustifolia	Mediterranean woodland	5.38 ± 0.93
Phoenix dactylifera	Hot desert	1.11 ± 0.24
Pistacia lentiscus	Mediterranean woodland	2.49 ± 0.41
Quercus coccifera	Mediterranean woodland	3.09 ± 1.04
Senna italica	Hot desert	3.03 ± 0.64
Smilax aspera	Mediterranean woodland	0.83 ± 0.15
Smilax japicanga	Tropical dry forest	0.46 ± 0.24

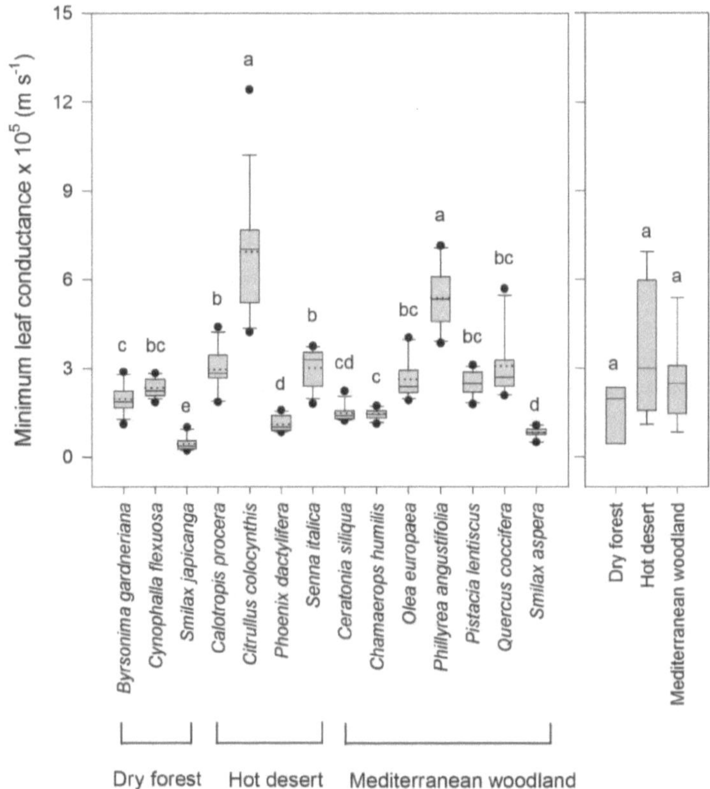

Fig. 14. Left, minimum leaf conductances (g_{min}) of 14 xerophilous plant species (n ≥ 9 biological replicates). Right, g_{min} grouped according to the original plant habitat (three dry forest species, four desert species, and seven Mediterranean species). Different letters indicate a significant difference (p < 0.05) of the g_{min}.

3.2.2 Leaf photosynthetic thermal tolerance

We compared the critical temperature (T_{crit}) of 14 flowering plant species for providing insight into the leaf photosynthetic thermal tolerance within the three biomes. Dark-adapted fully developed leaves of xerophilous plant species were exposed to heat increasing at 1°C min⁻¹, and the maximum quantum yield of photosystem II ($F_v F_m^{-1}$) was measured at every 2.5 min without recovery time. The T_{crit} of xerophilous plants ranged from 42.6 ± 1.0°C (mean ±

standard deviation) for *C. colocynthis* (a desert plant) to 49.3 ± 2.8°C for *S. japicanga* (a dry forest plant). For more details, see Table 9. T_{crit} varied among species by more than 6°C, but no difference (F (2) = 1.64, p = 0.239) between habitats was identified (Fig. 15).

Table 9. Critical temperatures (T_{crit}) of 14 flowering plant species from three hot and dry biomes. Each value represents the mean ± standard deviation ($n \geq 10$ biological replicates).

Plant species	Original habitat	T_{crit}
Byrsonima gardneriana	Tropical dry forest	45.8 ± 1.6
Calotropis procera	Hot desert	44.1 ± 3.2
Ceratonia siliqua	Mediterranean woodland	44.2 ± 1.2
Chamaerops humilis	Mediterranean woodland	46.8 ± 1.2
Citrullus colocynthis	Hot desert	42.6 ± 1.0
Cynophalla flexuosa	Tropical dry forest	43.5 ± 3.1
Olea europaea	Mediterranean woodland	43.1 ± 0.9
Phillyrea angustifolia	Mediterranean woodland	42.9 ± 0.7
Phoenix dactylifera	Hot desert	44.9 ± 0.8
Pistacia lentiscus	Mediterranean woodland	44.1 ± 1.1
Quercus coccifera	Mediterranean woodland	44.0 ± 3.3
Senna italica	Hot desert	43.3 ± 1.8
Smilax aspera	Mediterranean woodland	47.0 ± 1.5
Smilax japicanga	Tropical dry forest	49.3 ± 2.8

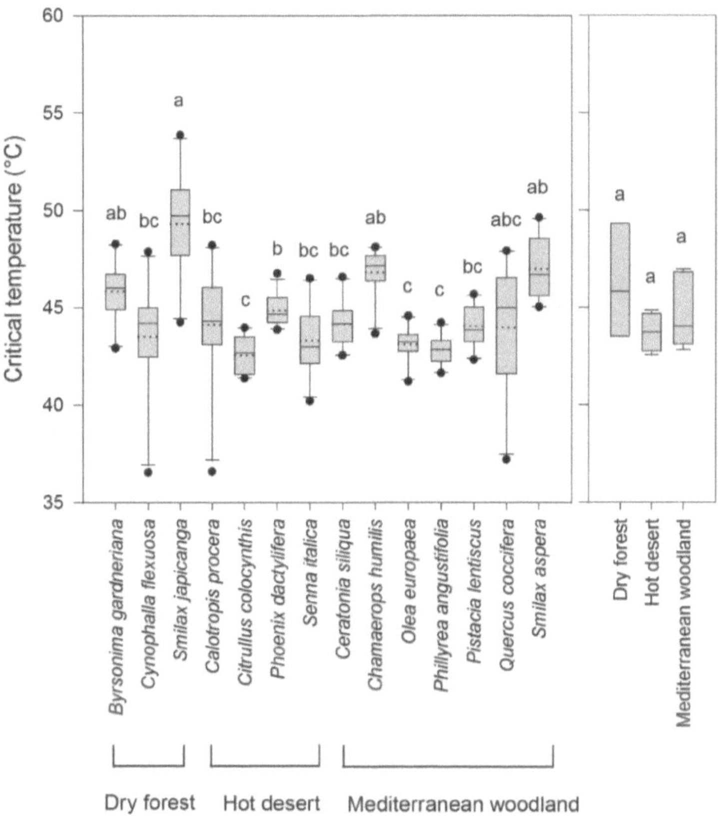

Fig. 15. Left, critical temperatures (T_{crit}) of 14 xerophilous plant species (n $\geq$ 10 biological replicates). Right, T_{crit} arranged according to the original plant habitat (three dry forest species, four desert species, and seven Mediterranean species). Different letters indicate a significant difference (p < 0.05) of the T_{crit}.

3.2.3 Leaf photosynthetic thermal tolerance and cuticular transpiration barrier

Minimum leaf conductance (g_{min}) was negatively correlated with leaf critical temperature, i. e. the lower the minimum conductance, the higher the critical temperature (Fig. 16). The equation $g_{min} = (-0.6715 \cdot T_{crit}) + 32.591$ determines the relationship between g_{min} and T_{crit}, thereby indicating a link between leaf photosynthetic thermal tolerance and cuticular transpiration barrier.

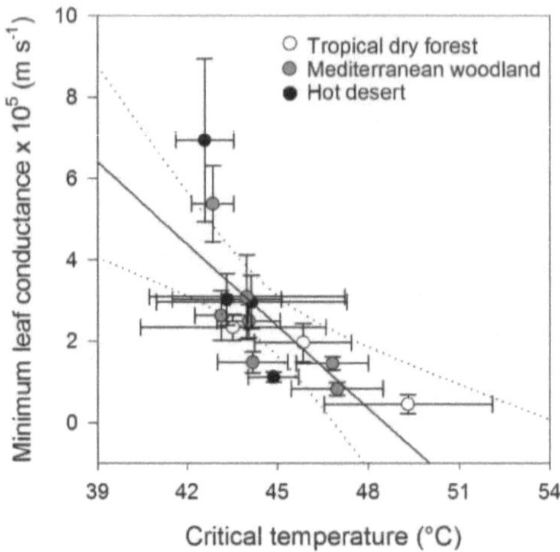

Fig. 16. Minimum leaf conductances of 14 flowering plant species from three hot and dry biomes decrease with increasing critical temperatures (R^2 = 0.53, p = 0.002). Each point corresponds to one plant species (mean ± standard deviation, n ≥ 9 biological replicates). Dashed lines represent 95% confidence intervals.

3.2.4 Chemical analysis of the cuticular leaf waxes

The cuticular waxes primarily make up the cuticular transpiration barrier. We investigated qualitatively and quantitatively the chemical composition of the cuticular leaf waxes of 14 flowering plant species from three hot and dry biomes using gas chromatography. The xerophilous plants examined displayed a wide range of wax content, wherein the desert plant *Citrullus colocynthis* had the lowest total wax (4.20 ± 0.44 µg cm^{-2}) and the Mediterranean *Phillyrea angustifolia* exhibited the highest (310.70 ± 3.84 µg cm^{-2}). Below, we presented the individual cuticular wax composition of the studied plant species (in alphabetical order), and we discussed the relationship between the wax chemical composition and the transpiration barrier of xerophilous flowering plants in the general discussion section (page 105).

3.2.5 Cuticular leaf wax of Byrsonima gardneriana (tropical dry forest)

Cuticular leaf wax of *B. gardneriana* was qualitatively and quantitatively analysed. More than 93.5% of the wax extract was identified using GC-MS. Total wax coverage of fully expanded leaves of *B. gardneriana* amounted to 24.13 ± 3.17 µg cm^{-2} (mean ± standard deviation, n = 5 biological replicates). The leaf wax comprised a major fraction of very-long-chain acyclic (VLC) component classes (60.2% of the total wax, and wax coverage of 14.51 ± 3.16 µg cm^{-2}) and a minor fraction of cyclic components (33.8%, 8.16 ± 0.48 µg cm^{-2}).

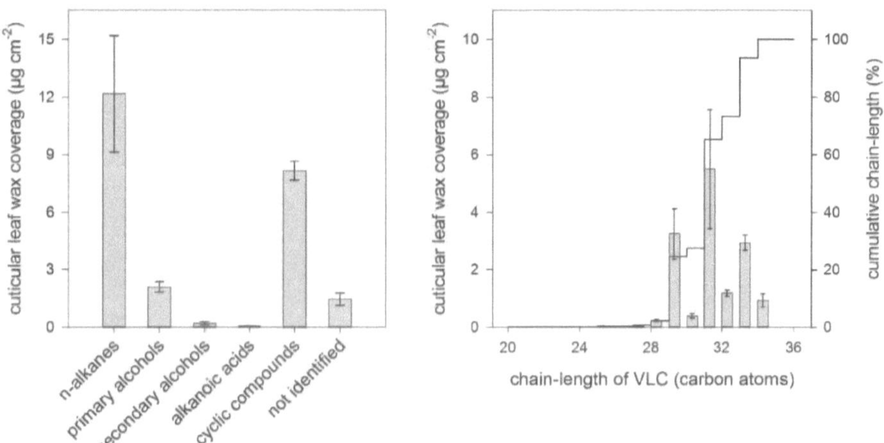

Fig. 17. Cuticular wax composition of *Byrsonima gardneriana* leaves from the tropical dry forest, northeastern Brazil. Left: cuticular wax coverage arranged according to the component classes, and right: the carbon chain-length distribution of the very-long-chain aliphatic (VLC) components and their cumulative contribution to the total VLC fraction (mean ± standard deviation, n = 5 biological replications).

N-alkanes were the most prominent VLC component class (50.4%, 12.17 ± 3.03 µg cm^{-2}), followed by primary alcohols (8.7%, 2.09 ± 0.26 µg cm^{-2}). Aditional VLC component classes detected in the cuticular wax were alkanoic acids and secondary alcohols. Carbon chain-lengths ranged from C_{20} to C_{34}, and the predominant chain-lengths were C_{29},

C_{31}, and C_{33} (Fig. 17). *N*-hentriacontane dominated the VLC components (22.2%, 5.35 ± 2.09 µg cm^{-2}). The average carbon chain-length (ACL) of the VLC wax components was 31.12 carbon atoms. The main cyclic components were alpha-amyrin and beta-amyrin with a cuticular wax coverage of 2.75 ± 0.26 µg cm^{-2} (11.4%) and 2.04 ± 0.27 µg cm^{-2} (8.5%), respectively (Table 10).

Table 10. Cuticular wax composition of *Byrsonima japicanga* leaves from the tropical dry forest, northeastern Brazil (mean ± standard deviation, n = 5 biological replications).

Compound class	Carbon chain-length	Wax coverage (µg cm^{-2})
alkanoic acids	20	0.008 ± 0.001
	32	0.053 ± 0.010
primary alcohols	31	0.083 ± 0.017
	32	0.895 ± 0.072
	33	0.180 ± 0.024
	34	0.933 ± 0.228
n-alkanes	25	0.038 ± 0.008
	27	0.048 ± 0.007
	28	0.238 ± 0.034
	29	3.172 ± 0.910
	30	0.391 ± 0.076
	31	5.347 ± 2.087
	32	0.178 ± 0.044
	33	2.757 ± 0.281
secondary alcohols	29	0.076 ± 0.050
	31	0.063 ± 0.029
	32	0.053 ± 0.010
total very-long-chain aliphatics		*14.514 ± 3.162*
alpha-amyrin		2.746 ± 0.257
beta-amyrin		2.038 ± 0.266
betulinic acid		0.207 ± 0.180
delta-amyrin		0.140 ± 0.072
epi-fridelanol		0.599 ± 0.052
erythrodiol		0.055 ± 0.008
fridelin		0.261 ± 0.093
lupeol		1.639 ± 0.451
oleanolic acid		0.220 ± 0.099
ursolic acid		0.102 ± 0.056
uvaol		0.050 ± 0.011
total cyclic aliphatics		*8.058 ± 0.488*
beta-tocopherol		0.099 ± 0.033
total cyclic aromatics		*0.099 ± 0.033*

Compound class	Carbon chain-length	Wax coverage (μg cm^{-2})
total cyclics		*8.157 ± 0.483*
not identified		1.459 ± 0.308
total wax		*24.130 ± 3.166*

3.2.6 Cuticular leaf wax of Calotropis procera (hot desert)

Cuticular leaf wax of *C. procera* was qualitatively and quantitatively analysed. More than 86% of the wax extract was identified using GC-MS. Total wax coverage of fully expanded leaves of *C. procera* amounted to 11.28 ± 1.33 μg cm^{-2} (mean ± standard deviation, n = 5 biological replicates). The leaf wax comprised a major fraction of very-long-chain acyclic (VLC) component classes (54.4% of the total wax, and wax coverage of 6.14 ± 1.08 μg cm^{-2}) and a minor fraction of cyclic components (31.7%, 3.57 ± 0.34 μg cm^{-2}).

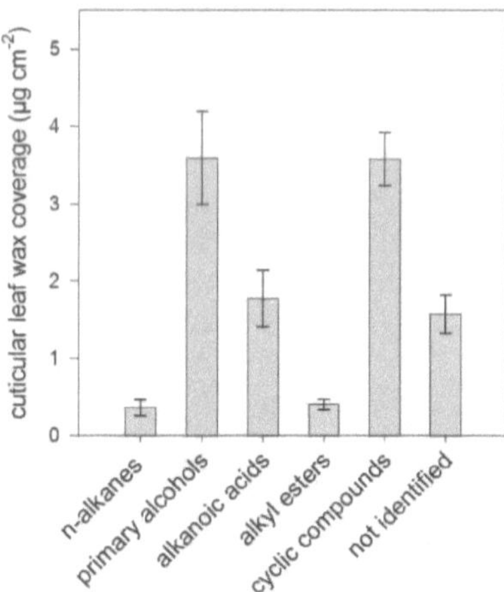

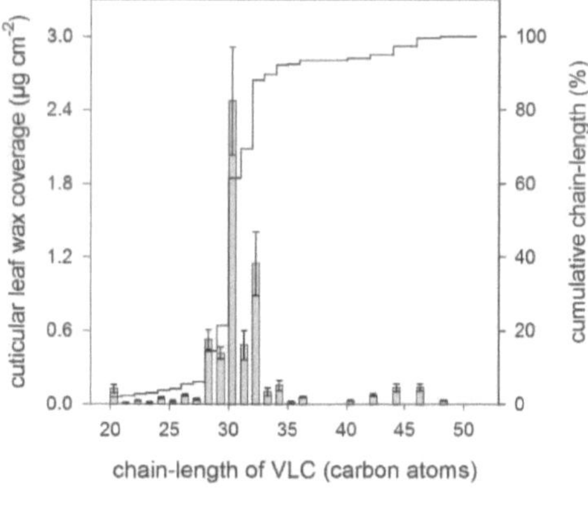

Fig. 18. Cuticular wax composition of *Calotropis procera* leaves original from the hot desert, northeastern Saudi Arabia. Left: cuticular wax coverage arranged according to the component classes, and right: the carbon chain-length distribution of the very-long-chain aliphatic (VLC) components and their cumulative contribution to the total VLC fraction (mean ± standard deviation, n = 5 biological replications).

Primary alcohols were the most prominent VLC component class (31.8%, 3.59 ± 0.59 µg cm^{-2}), followed by alkanoic acids (15.7%, 1.77 ± 0.36 µg cm^{-2}). Aditional VLC component classes detected in the cuticular wax were n-alkanes and alkyl esters. Carbon chain-lengths ranged from C$_{20}$ to C$_{48}$, and the predominant chain-lengths were C$_{28}$, C$_{30}$, and C$_{32}$ (Fig. 18). Triacontanol dominated the VLC components (15.2%, 1.72 ± 0.30 µg cm^{-2}). The average carbon chain-length (ACL) of the VLC wax components was 30.97 carbon atoms. The main cyclic components were delta-amyrin and alpha-amyrin with a cuticular wax coverage of 1.90 ± 0.26 µg cm^{-2} (16.9%) and 0.72 ± 0.08 µg cm^{-2} (6.4%), respectively (Table 11).

Table 11. Cuticular wax composition of *Calotropis procera* leaves original from the hot desert, northeastern Saudi Arabia. (mean ± standard deviation, n = 5 biological replications).

Compound class	Carbon chain-length	Wax coverage (µg cm^{-2})
	20	0.105 ± 0.026
	21	0.009 ± 0.004
	22	0.023 ± 0.002
	23	0.013 ± 0.005
	24	0.051 ± 0.008
	25	0.011 ± 0.004
	26	0.037 ± 0.006
alkanoic acids	27	0.016 ± 0.001
	28	0.109 ± 0.020
	29	0.096 ± 0.022
	30	0.751 ± 0.143
	31	0.099 ± 0.025
	32	0.403 ± 0.107
	33	0.033 ± 0.019
	34	0.025 ± 0.014
	20	0.022 ± 0.006
	22	0.008 ± 0.002
	26	0.036 ± 0.007
	27	0.020 ± 0.005
	28	0.415 ± 0.061
primary alcohols	29	0.210 ± 0.027
	30	1.720 ± 0.298
	31	0.144 ± 0.017
	32	0.739 ± 0.154
	33	0.075 ± 0.020
	34	0.128 ± 0.027

Compound class	Carbon chain-length	Wax coverage ($\mu g\ cm^{-2}$)
	35	0.020 ± 0.005
	36	0.058 ± 0.007
alkyl esters	25	0.011 ± 0.005
	27	0.008 ± 0.004
	29	0.109 ± 0.023
	31	0.237 ± 0.082
n-alkanes	25	0.011 ± 0.005
	27	0.008 ± 0.004
	29	0.109 ± 0.023
	31	0.237 ± 0.082
total very-long-chain aliphatics		*6.136 ± 1.083*
alpha-amyrin		0.722 ± 0.080
amyrin derivative		0.483 ± 0.065
amyrin derivative		0.241 ± 0.027
beta-amyrin		0.116 ± 0.013
betulinic acid		0.049 ± 0.019
delta-amyrin		1.903 ± 0.257
lupeol		0.060 ± 0.009
total cyclic aliphatics		*3.574 ± 0.340*
total cyclics		*3.574 ± 0.340*
not identified		1.572 ± 0.246
total wax		*11.283 ± 1.325*

3.2.7 Cuticular leaf wax of Ceratonia siliqua (Mediterranean woodland)

Cuticular leaf wax of *C. siliqua* was qualitatively and quantitatively analysed. More than 94% of the wax extract was identified using GC-MS. Total wax coverage of fully expanded leaves of *C. siliqua* amounted to 36.06 ± 2.72 $\mu g\ cm^{-2}$ (mean ± standard deviation, n = 4 biological replicates). The leaf wax comprised a major fraction of very-long-chain acyclic (VLC) component classes (60.2% of the total wax, and wax coverage of 21.70 ± 1.90 $\mu g\ cm^{-2}$) and a minor fraction of cyclic components (34.0%, 12.25 ± 1.11 $\mu g\ cm^{-2}$).

Alkanol acetates were the most prominent VLC component class (33.5%, 12.07 ± 2.25 $\mu g\ cm^{-2}$), followed by primary alcohols (20.6%, 7.44 ± 2.05 $\mu g\ cm^{-2}$). Aditional VLC component class detected in the cuticular wax were *n*-alkanes, aldehydes, and alkyl esters. Carbon chain-lengths ranged from C_{22} to C_{50}, and the predominant chain-lengths were C_{26}, C_{28}, and C_{29} (Fig. 19).

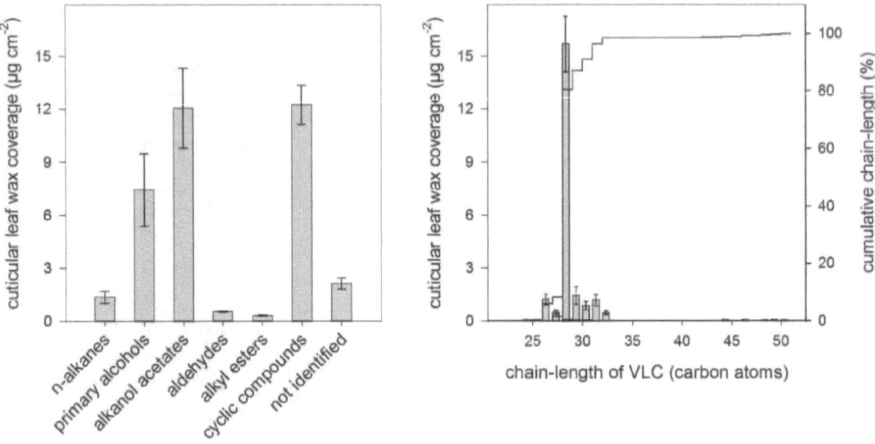

Fig. 19. Cuticular wax composition of *Ceratonia siliqua* leaves from the Mediterranean woodland, southern Portugal. Left: cuticular wax coverage arranged according to the component classes, and right: the carbon chain-length distribution of the very-long-chain aliphatic (VLC) components and their cumulative contribution to the total VLC fraction (mean ± standard deviation, n = 4 biological replications).

Hexacosyl acetate dominated the VLC components (30.3%, 10.92 ± 2.06 µg cm^{-2}). The average carbon chain-length (ACL) of the VLC wax components was 28.51 carbon atoms. The main cyclic components were lupeol and germanicol with a cuticular wax coverage of 3.32 ± 0.44 µg cm^{-2} (9.2%) and 2.21 ± 0.46 µg cm^{-2} (6.1%), respectively (Table 12).

Table 12. Cuticular wax composition of *Ceratonia siliqua* leaves from the Mediterranean woodland, southern Portugal (mean ± standard deviation, n = 4 biological replications).

Compound class	Carbon chain-length	Wax coverage (µg cm^{-2})
	22	0.007 ± 0.001
	24	0.048 ± 0.019
	25	0.014 ± 0.003
primary alcohols	26	0.578 ± 0.335
	27	0.167 ± 0.053
	28	4.590 ± 1.517

Compound class	Carbon chain-length	Wax coverage (μg cm^{-2})		
	29	1.103	±	0.453
	30	0.602	±	0.260
	31	0.262	±	0.086
	32	0.073	±	0.028
	42	0.012	±	0.003
	43	0.004	±	0.001
	44	0.062	±	0.002
	45	0.007	±	0.000
alkyl esters	46	0.062	±	0.004
	47	0.009	±	0.000
	48	0.044	±	0.006
	49	0.057	±	0.005
	50	0.049	±	0.005
	26	0.023	±	0.006
aldehydes	28	0.174	±	0.044
	32	0.337	±	0.061
	26	0.619	±	0.235
	27	0.208	±	0.073
alkanol acetates	28	10.923	±	2.057
	30	0.264	±	0.066
	32	0.051	±	0.009
	25	0.014	±	0.002
n-alkanes	27	0.100	±	0.010
	29	0.323	±	0.050
	31	0.912	±	0.310
total very-long-chain aliphatics		*21.699*	*±*	*1.897*
beta-amyrin		1.501	±	0.360
germanicol		2.205	±	0.463
germanicone		0.632	±	0.321
lupenon		1.143	±	0.312
lupeol		3.322	±	0.435
lupeol derivative 1		0.615	±	0.190
lupeol derivative 2		1.445	±	0.457
triterpenoid 1		0.672	±	0.223
triterpenoid 2		0.315	±	0.044
triterpenoid 3		0.112	±	0.024
triterpenoid 4		0.156	±	0.031
triterpenoid 5		0.129	±	0.030
total cyclic aliphatics		*12.247*	*±*	*1.111*
total cyclics		*12.247*	*±*	*1.111*
not identified		2.116	±	0.321
total wax		*36.062*	*±*	*2.724*

3.2.8 Cuticular leaf wax of Chamaerops humilis (Mediterranean woodland)

Cuticular leaf wax of *C. humilis* was qualitatively and quantitatively analysed. More than 88% of the wax extract was identified using GC-MS. Total wax coverage of fully expanded leaves of *C. humilis* amounted to 26.99 ± 1.15 µg cm^{-2} (mean ± standard deviation, n = 4 biological replicates). The leaf wax comprised very-long-chain acyclic (VLC) component classes (88.8% of the total wax, and wax coverage of 23.98 ± 1.11 µg cm^{-2}) and cyclic components were absent.

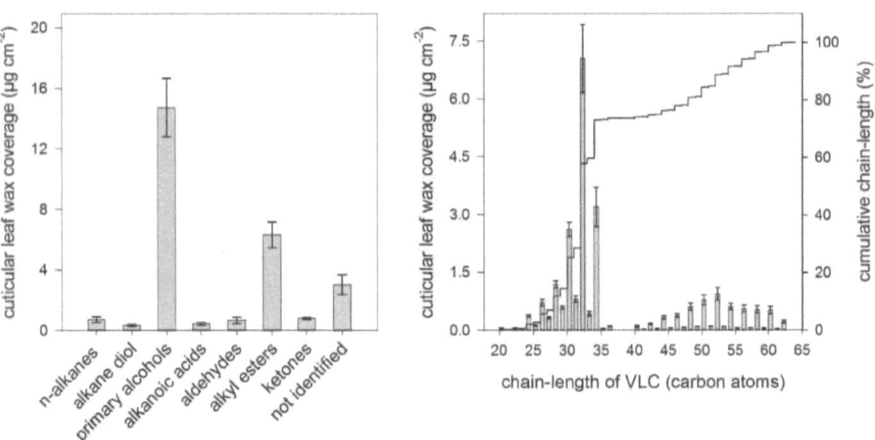

Fig. 20. Cuticular wax composition of *Chamaerops humilis* leaves from the Mediterranean woodland, southern Portugal. Left: cuticular wax coverage arranged according to the component classes, and right: the carbon chain-length distribution of the very-long-chain aliphatic (VLC) components and their cumulative contribution to the total VLC fraction (mean ± standard deviation, n = 4 biological replications).

Primary alcohols were the most prominent VLC component class (54.5%, 14.72 ± 1.92 µg cm^{-2}), followed by alkyl esters (23.4%, 6.32 ± 0.84 µg cm^{-2}). Aditional VLC component classes detected in the cuticular wax were alkane diols, alkanoic acids, aldehydes, and ketones. Carbon chain-lengths ranged from C_{20} to C_{62}, and the predominant chain-lengths were C_{30},

C$_{32}$, and C$_{34}$ (Fig. 20). Dotriacontanol dominated the VLC components (25.4%, 6.85 ± 1.04

µg cm^{-2}, Table 13). The average carbon chain-length (ACL) of the VLC wax components was

36.62 carbon atoms.

Table 13. Cuticular wax composition of *Chamaerops humilis* leaves from the Mediterranean

woodland, southern Portugal (mean ± standard deviation, n = 4 biological replications).

Compound class	Carbon chain-length	Wax coverage (µg cm^{-2})
	20	0.020 ± 0.012
	21	0.005 ± 0.003
	22	0.011 ± 0.001
	23	0.006 ± 0.001
	24	0.057 ± 0.025
	25	0.012 ± 0.003
alkanoic acids	26	0.061 ± 0.018
	27	0.028 ± 0.004
	28	0.112 ± 0.049
	30	0.064 ± 0.026
	32	0.038 ± 0.019
	34	0.028 ± 0.006
	36	0.005 ± 0.001
	20	0.016 ± 0.003
	21	0.006 ± 0.005
	22	0.023 ± 0.005
	23	0.017 ± 0.005
	24	0.249 ± 0.017
	25	0.041 ± 0.009
	26	0.527 ± 0.066
	27	0.070 ± 0.012
primary alcohols	28	0.837 ± 0.115
	29	0.154 ± 0.027
	30	2.153 ± 0.279
	31	0.271 ± 0.080
	32	6.850 ± 1.035
	33	0.270 ± 0.066
	34	3.125 ± 0.579
	35	0.038 ± 0.009
	36	0.073 ± 0.010
	40	0.093 ± 0.022
	41	0.021 ± 0.003
alkyl esters	42	0.158 ± 0.027
	43	0.038 ± 0.003
	44	0.331 ± 0.057

Compound class	Carbon chain-length	Wax coverage ($\mu g\ cm^{-2}$)
	45	0.052 ± 0.004
	46	0.376 ± 0.052
	47	0.066 ± 0.003
	48	0.607 ± 0.106
	49	0.085 ± 0.001
	50	0.782 ± 0.145
	51	0.094 ± 0.004
	52	0.937 ± 0.188
	53	0.081 ± 0.006
	54	0.607 ± 0.091
	55	0.051 ± 0.005
	56	0.551 ± 0.106
	57	0.053 ± 0.005
	58	0.531 ± 0.106
	59	0.042 ± 0.008
	60	0.517 ± 0.113
	61	0.026 ± 0.004
	62	0.218 ± 0.040
	26	0.036 ± 0.014
	28	0.117 ± 0.062
aldehydes	30	0.337 ± 0.097
	32	0.135 ± 0.055
	34	0.026 ± 0.014
	36	0.011 ± 0.008
	25	0.014 ± 0.007
	27	0.124 ± 0.019
ketones	29	0.269 ± 0.022
	31	0.262 ± 0.023
	33	0.106 ± 0.013
	20	0.009 ± 0.006
	22	0.024 ± 0.004
	24	0.063 ± 0.011
	26	0.060 ± 0.014
alkane diols	28	0.108 ± 0.048
	30	0.037 ± 0.011
	32	0.018 ± 0.007
	34	0.016 ± 0.005
	36	0.008 ± 0.002
	25	0.055 ± 0.021
	26	0.028 ± 0.021
_n_-alkanes	27	0.101 ± 0.027
	28	0.010 ± 0.002
	29	0.168 ± 0.047
	30	0.026 ± 0.011

Compound class	Carbon chain-length	Wax coverage ($\mu g\ cm^{-2}$)
	31	0.273 ± 0.094
	32	0.007 ± 0.003
	33	0.051 ± 0.013
total very-long-chain aliphatics		*23.978 ± 1.105*
not identified		3.013 ± 0.769
total wax		*26.991 ± 1.322*

3.2.9 Cuticular leaf wax of Citrullus colocynthis (hot desert)

The cuticular wax composition of *C. colocynthis* leaves was presented in chapter one (see page 49), and wax particularities were discussed in relation to the life strategy adopted by this plant species to deal with water and heat stresses.

3.2.10 Cuticular leaf wax of Cynophalla flexuosa (tropical dry forest)

Cuticular leaf wax of *C. flexuosa* was qualitatively and quantitatively analysed. More than 87% of the wax extract was identified using GC-MS. Total wax coverage of fully expanded leaves of *C. flexuosa* amounted to 17.04 ± 2.20 $\mu g\ cm^{-2}$ (mean ± standard deviation, n = 5 biological replicates). The leaf wax comprised a major fraction of very-long-chain acyclic (VLC) component classes (59.6% of the total wax, and wax coverage of 10.15 ± 1.54 $\mu g\ cm^{-2}$) and a minor fraction of cyclic components (27.6%, 4.71 ± 0.24 $\mu g\ cm^{-2}$).

N-alkanes were the most prominent VLC component class (29.3%, 4.98 ± 0.49 $\mu g\ cm^{-2}$), followed by alkanoic acids (18.4%, 3.14 ± 0.64 $\mu g\ cm^{-2}$). Aditional VLC component classes detected in the cuticular wax were primary alcohols and alkyl esters. Carbon chain-lengths ranged from C_{20} to C_{50}, and the predominant chain-lengths were C_{30}, C_{31}, and C_{33} (Fig. 21).

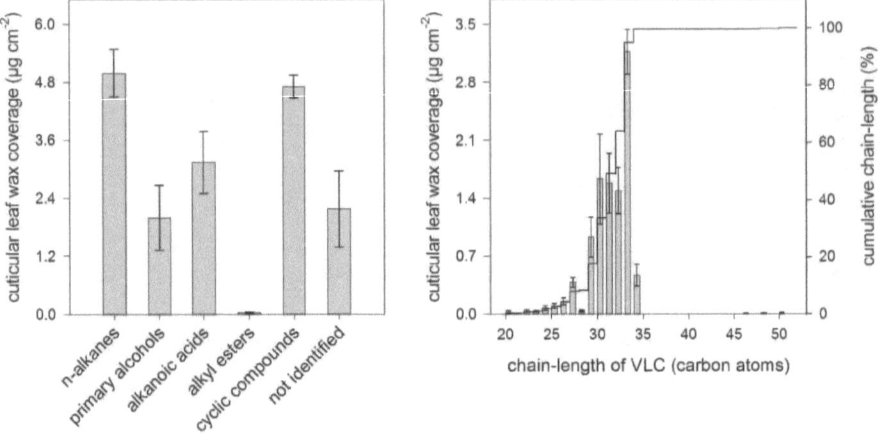

Fig. 21. Cuticular wax composition of *Cynophalla flexuosa* leaves from the tropical dry forest, northeastern Brazil. Left: cuticular wax coverage arranged according to the component classes, and right: the carbon chain-length distribution of the very-long-chain aliphatic (VLC) components and their cumulative contribution to the total VLC fraction (mean ± standard deviation, n = 5 biological replications).

N-tritriacontane dominated the VLC components (17.8%, 3.03 ± 0.28 µg cm⁻²). The average carbon chain-length (ACL) of the VLC wax components was 31.18 carbon atoms. The main cyclic components were lupeol and betulin with a cuticular wax coverage of 2.21 ± 0.20 µg cm⁻² (12.9%) and 0.49 ± 0.18 µg cm⁻² (2.9%), respectively (Table 14).

Table 14. Cuticular wax composition of *Cynophalla flexuosa* leaves from the tropical dry forest, northeastern Brazil (mean ± standard deviation, n = 5 biological replications).

Compound class	Carbon chain-length	Wax coverage (µg cm⁻²)
	20	0.030 ± 0.014
	21	0.010 ± 0.004
alkanoic acids	22	0.023 ± 0.010
	23	0.032 ± 0.011
	24	0.071 ± 0.027
	25	0.045 ± 0.012

Compound class	Carbon chain-length	Wax coverage ($\mu g\ cm^{-2}$)
	26	0.091 ± 0.024
	27	0.301 ± 0.042
	28	0.037 ± 0.011
	29	0.687 ± 0.209
	30	0.702 ± 0.153
	31	0.286 ± 0.087
	32	0.502 ± 0.108
	33	0.058 ± 0.022
	34	0.262 ± 0.056
	22	0.014 ± 0.005
	30	0.928 ± 0.395
primary alcohols	31	0.299 ± 0.105
	32	0.470 ± 0.155
	33	0.077 ± 0.022
	34	0.208 ± 0.084
	46	0.007 ± 0.002
alkyl esters	48	0.013 ± 0.003
	50	0.016 ± 0.004
	25	0.052 ± 0.019
	26	0.062 ± 0.023
	27	0.081 ± 0.034
n-alkanes	29	0.242 ± 0.080
	31	0.995 ± 0.213
	32	0.519 ± 0.133
	33	3.034 ± 0.276
total very-long-chain aliphatics		*10.154 ± 1.542*
alpha-amyrin		0.029 ± 0.003
beta-amyrin		0.319 ± 0.101
betulin		0.495 ± 0.184
lupenon		0.187 ± 0.079
lupeol		2.207 ± 0.198
lupeol derivative		0.366 ± 0.160
total cyclic aliphatics		*4.212 ± 0.229*
alpha-tocopherol		0.354 ± 0.076
beta-tocopherol		0.121 ± 0.012
delta-tocopherol		0.019 ± 0.005
total cyclic aromatics		*0.494 ± 0.081*
total cyclics		*4.705 ± 0.237*
not identified		2.177 ± 0.785
total wax		*17.036 ± 2.202*

3.2.11 Cuticular leaf wax of Olea europaea var. sylvestris (Mediterranean woodland)

Cuticular leaf wax of *O. europaea* was qualitatively and quantitatively analysed. More than 96% of the wax extract was identified using GC-MS. Total wax coverage of fully expanded leaves of *O. europaea* amounted to 145.65 ± 10.41 µg cm^{-2} (mean ± standard deviation, n = 4 biological replicates). The leaf wax comprised a major fraction of cyclic components (88.6% of the total wax, and wax coverage of 129.14 ± 9.11 µg cm^{-2}) and a minor fraction of very-long-chain acyclic (VLC) component classes (8.3%, 12.05 ± 0.93 µg cm^{-2}).

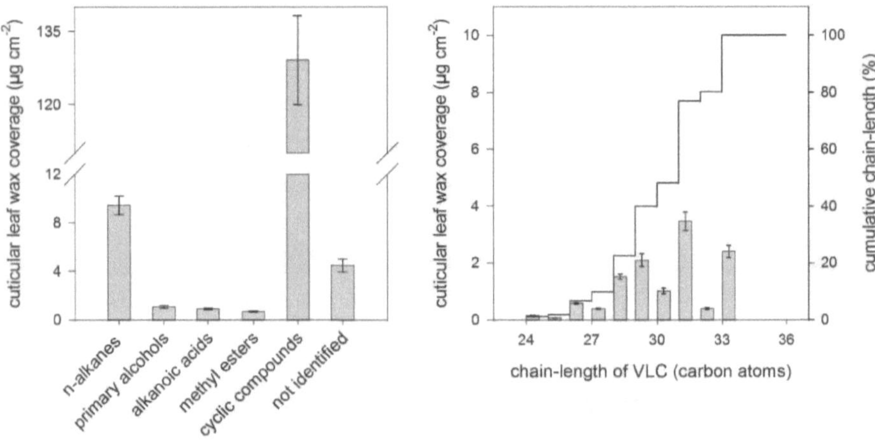

Fig. 22. Cuticular wax composition of *Olea europaea var. sylvestris* leaves from the Mediterranean woodland, southern Portugal. Left: cuticular wax coverage arranged according to the component classes, and right: the carbon chain-length distribution of the very-long-chain aliphatic (VLC) components and their cumulative contribution to the total VLC fraction (mean ± standard deviation, n = 4 biological replications).

The main cyclic components were oleanolic acid and ursolic acid with a cuticular wax coverage of 77.78 ± 9.86 µg cm^{-2} (53.4%) and 15.15 ± 1.21 µg cm^{-2} (10.4%), respectively (Table 15). *N*-alkanes were the most prominent VLC component class (6.5%, 9.43 ± 0.76 µg cm^{-2}), followed by primary alcohols (0.7%, 1.07 ± 0.11 µg cm^{-2}). Aditional VLC component classes detected in the cuticular wax were alkanoic acids and methyl esters. Carbon chain-lengths ranged from C_{24} to C_{33}, and the predominant chain-lengths were C_{29}, C_{31}, and C_{33} (Fig. 22). *N*-hentriacontane dominated the VLC components (2.4%, 3.47 ± 0.36 µg cm^{-2}). The average carbon chain-length (ACL) of the VLC wax components was 30.13 carbon atoms.

Table 15. Cuticular wax composition of *Olea europaea var. sylvestris* leaves from the Mediterranean woodland, southern Portugal (mean ± standard deviation, n = 4 biological replications).

Compound class	Carbon chain-length	Wax coverage (µg cm^{-2})
alkanoic acids	24	0.036 ± 0.008
	26	0.079 ± 0.021
	28	0.615 ± 0.051
	30	0.164 ± 0.089
primary alcohols	24	0.102 ± 0.022
	26	0.212 ± 0.031
	28	0.389 ± 0.041
	30	0.365 ± 0.049
methyl esters	24	0.009 ± 0.009
	26	0.215 ± 0.007
	28	0.307 ± 0.017
	30	0.132 ± 0.030
n-alkanes	25	0.070 ± 0.010
	26	0.079 ± 0.019
	27	0.391 ± 0.028
	28	0.204 ± 0.017
	29	2.088 ± 0.259
	30	0.348 ± 0.019
	31	3.465 ± 0.363
	32	0.390 ± 0.048
	33	2.397 ± 0.243
total very-long-chain aliphatics		*12.054 ± 0.935*
betulinic acid		0.976 ± 0.126
alpha-amyrin		0.030 ± 0.028
uvaol		13.455 ± 1.534
ursolic acid		15.148 ± 1.209
beta-amyrin		0.531 ± 0.082
erythrodiol		12.398 ± 0.380
oleanolic acid		77.778 ± 9.858
hederagenin		5.564 ± 0.988
delta-amyrin		0.214 ± 0.023
triterpenoid 1		1.723 ± 0.743

Compound class	Carbon chain-length	Wax coverage ($\mu g\ cm^{-2}$)
triterpenoid 2		1.318 ± 0.204
total cyclic aliphatics		*129.136 ± 9.107*
total cyclics		*129.136 ± 9.107*
not identified		4.461 ± 0.613
total wax		*145.650 ± 12.021*

3.2.12 Cuticular leaf wax of Phillyrea angustifolia (Mediterranean woodland)

Cuticular leaf wax of *P. angustifolia* was qualitatively and quantitatively analysed. More than 94% of the wax extract was identified using GC-MS. Total wax coverage of fully expanded leaves of *P. angustifolia* amounted to 310.70 ± 3.84 µg cm^{-2} (mean ± standard deviation, n = 4 biological replicates). The leaf wax comprised a major fraction of cyclic components (89.2% of the total wax, and wax coverage of 277.15 ± 5.26 µg cm^{-2}) and a minor fraction of very-long-chain acyclic (VLC) component classes (5.4%, 16.71 ± 1.06 µg cm^{-2}).

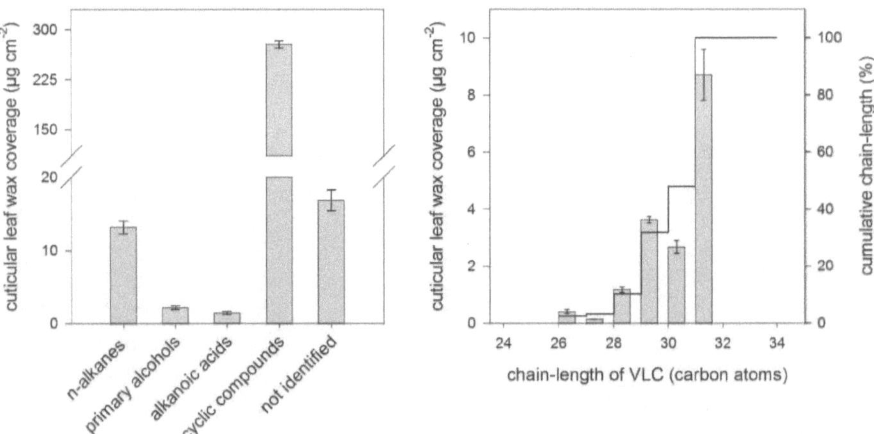

Fig. 23. Cuticular wax composition of *Phillyrea angustifolia* leaves from the Mediterranean woodland, southern Portugal. Left: cuticular wax coverage arranged according to the component classes, and right: the carbon chain-length distribution of the very-long-chain aliphatic (VLC) components and their cumulative contribution to the total VLC fraction (mean ± standard deviation, n = 4 biological replications).

The main cyclic components were oleanolic acid and ursolic acid with a cuticular wax coverage of 144.15 ± 3.04 µg cm^{-2} (46.4%) and 91.65 ± 5.46 µg cm^{-2} (29.5%), respectively (Table 16). *N*-alkanes were the most prominent VLC component class (4.2%, 13.16 ± 0.86 µg cm^{-2}), followed by primary alcohols (0.7%, 2.15 ± 0.22 µg cm^{-2}). Aditional VLC component class detected in the cuticular wax was alkanoic acids. Carbon chain-lengths ranged from C_{26} to C_{31}, and the predominant chain-lengths were C_{29}, C_{30}, and C_{31} (Fig. 23). *N*-hentriacontane dominated the VLC components (2.8%, 8.70 ± 1.03 µg cm^{-2}). The average carbon chain-length (ACL) of the VLC wax components was 30.04 carbon atoms.

Table 16. Cuticular wax composition of *Phillyrea angustifolia* leaves from the Mediterranean woodland, southern Portugal (mean ± standard deviation, n = 4 biological replications).

Compound class	Carbon chain-length	Wax coverage (µg cm^{-2})
	26	0.230 ± 0.085
alkanoic acids	28	0.225 ± 0.063
	30	0.942 ± 0.209
	26	0.169 ± 0.007
primary alcohols	28	0.799 ± 0.074
	30	1.182 ± 0.223
	27	0.135 ± 0.011
	28	0.148 ± 0.023
n-alkanes	29	3.626 ± 0.122
	30	0.551 ± 0.036
	31	8.702 ± 1.027
total very-long-chain aliphatics		16.707 ± 1.059
lupeol		12.177 ± 0.535
betulinic acid		6.110 ± 0.397
alpha-amyrin		2.073 ± 0.231
uvaol		6.711 ± 1.665
ursolic acid		91.652 ± 5.459
beta-amyrin		4.013 ± 0.241
erythrodiol		3.658 ± 0.722
oleanolic acid		144.150 ± 3.035
delta amyrin		1.374 ± 0.150
triterpenoid 1		1.442 ± 0.208
triterpenoid 2		1.164 ± 0.086
triterpenoid 3		0.966 ± 0.127
triterpenoid 4		0.753 ± 0.301
triterpenoid 5		0.908 ± 0.294
total cyclic aliphatics		277.150 ± 5.259

Compound class	Carbon chain-length	Wax coverage (μg cm^{-2})
total cyclics		277.150 ± 5.259
not identified		16.839 ± 1.650
total wax		*310.696 ± 4.438*

3.2.13 Cuticular leaf wax of Phoenix dactylifera (hot desert)

The cuticular wax composition of *P. dactylifera* leaflets was presented in chapter one (see page 49), and wax particularities were discussed in relation to the life strategy adopted by this plant species to deal with water and heat stresses.

3.2.14 Cuticular leaf wax of Pistacia lentiscus (Mediterranean woodland)

Cuticular leaf wax of *P. lentiscus* was qualitatively and quantitatively analysed. More than 85% of the wax extract was identified using GC-MS. Total wax coverage of fully expanded leaves of *P. lentiscus* amounted to 8.67 ± 1.70 μg cm^{-2} (mean ± standard deviation, n = 5 biological replicates). The leaf wax comprised a major fraction of very-long-chain acyclic (VLC) component classes (54.1% of the total wax, and wax coverage of 4.69 ± 1.30 μg cm^{-2}) and a minor fraction of cyclic components (31.3%, 2.72 ± 0.34 μg cm^{-2}).

Alkanoic acids were the most prominent VLC component class (24.1%, 2.09 ± 0.79 μg cm^{-2}), followed by *n*-alkanes (18.6%, 1.61 ± 0.44 μg cm^{-2}). Aditional VLC component class detected in the cuticular wax were primary alcohols, alkyl esters, and methyl esters. Carbon chain-lengths ranged from C_{24} to C_{48}, and the predominant chain-lengths were C_{28}, C_{30}, and C_{31} (Fig. 24).

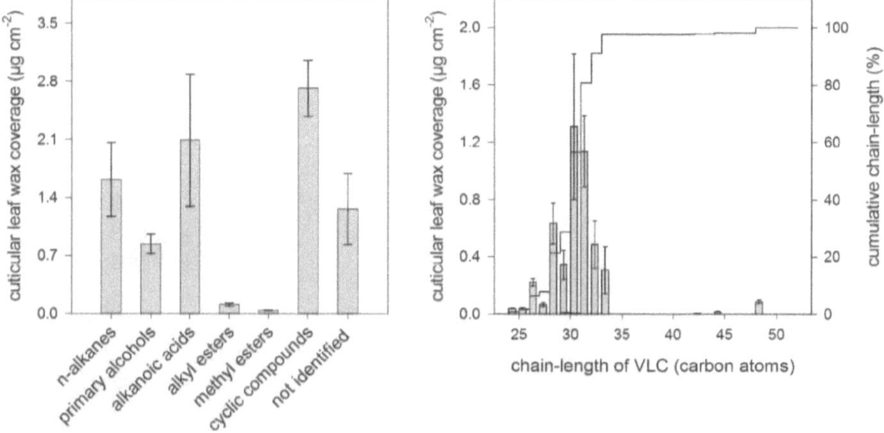

Fig. 24. Cuticular wax composition of *Pistacia lentiscus* leaves from the Mediterranean woodland, southern Portugal. Left: cuticular wax coverage arranged according to the component classes, and right: the carbon chain-length distribution of the very-long-chain aliphatic (VLC) components and their cumulative contribution to the total VLC fraction (mean ± standard deviation, n = 5 biological replications).

N-hentriacontane dominated the VLC components (10.8%, 0.94 ± 0.21 µg cm^{-2}). The average carbon chain-length (ACL) of the VLC wax components was 30.38 carbon atoms. The main cyclic components were lupeol and a lupeol derivative with a cuticular wax coverage of 1.85 ± 0.30 µg cm^{-2} (21.3%) and 0.38 ± 0.03 µg cm^{-2} (4.4%), respectively (Table 17).

Table 17. Cuticular wax composition of *Pistacia lentiscus* leaves from the Mediterranean woodland, southern Portugal (mean ± standard deviation, n = 5 biological replications).

Compound class	Carbon chain-length	Wax coverage (µg cm^{-2})
	24	0.039 ± 0.004
	25	0.021 ± 0.005
alkanoic acids	26	0.105 ± 0.022
	28	0.402 ± 0.134
	29	0.183 ± 0.059
	30	1.055 ± 0.463

Compound class	Carbon chain-length	Wax coverage ($\mu g\ cm^{-2}$)
	31	0.142 ± 0.030
	32	0.144 ± 0.094
	25	0.010 ± 0.004
	26	0.094 ± 0.016
	27	0.040 ± 0.007
	28	0.207 ± 0.027
primary alcohols	29	0.059 ± 0.019
	30	0.144 ± 0.018
	31	0.056 ± 0.009
	32	0.189 ± 0.032
	33	0.038 ± 0.007
	42	0.006 ± 0.002
alkyl esters	44	0.016 ± 0.003
	48	0.087 ± 0.013
methyl esters	26	0.015 ± 0.003
	28	0.024 ± 0.003
	25	0.008 ± 0.002
	26	0.009 ± 0.002
	27	0.027 ± 0.005
	29	0.101 ± 0.034
n-alkanes	30	0.110 ± 0.036
	31	0.939 ± 0.210
	32	0.152 ± 0.044
	33	0.268 ± 0.160
total very-long-chain aliphatics		*4.691 ± 1.301*
alpha-amyrin		0.025 ± 0.009
cholesterol		0.009 ± 0.003
lupenon		0.153 ± 0.050
lupeol		1.847 ± 0.298
lupeol derivative 1		0.377 ± 0.034
lupeol derivative 2		0.146 ± 0.038
oleanolic acid		0.070 ± 0.022
ursolic acid		0.089 ± 0.032
total cyclic aliphatics		*2.717 ± 0.337*
total cyclics		*2.717 ± 0.337*
not identified		1.263 ± 0.429
total wax		*8.670 ± 1.697*

3.2.15 Cuticular leaf wax of Quercus coccifera (Mediterranean woodland)

Cuticular leaf wax of *Q. coccifera* was qualitatively and quantitatively analysed. More than 90% of the wax extract was identified using GC-MS. Total wax coverage of fully expanded leaves of *Q. coccifera* amounted to 8.16 ± 1.57 µg cm^{-2} (mean ± standard deviation, n = 5 biological replicates). The leaf wax comprised a major fraction of cyclic components (45.8% of the total wax, and wax coverage of 3.73 ± 0.81 µg cm^{-2}) and a minor fraction of very-long-chain acyclic (VLC) component classes (44.4%, 3.63 ± 0.75 µg cm^{-2}).

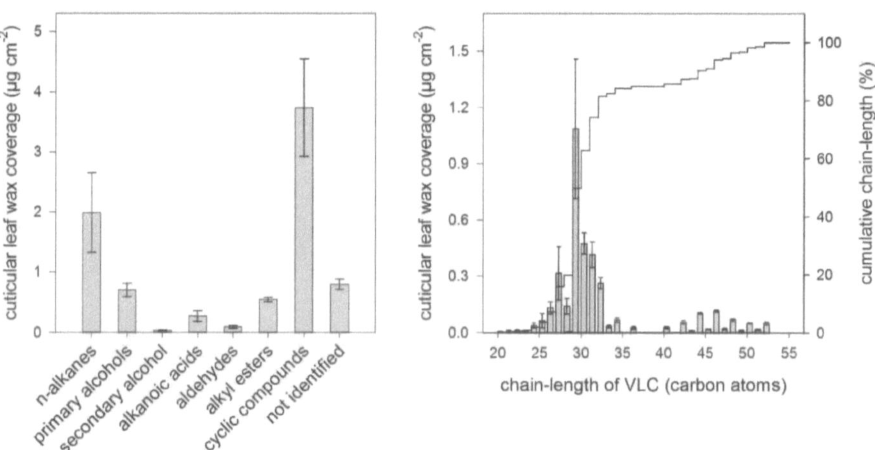

Fig. 25. Cuticular wax composition of *Quercus coccifera* leaves from the Mediterranean woodland, southern Portugal. Left: cuticular wax coverage arranged according to the component classes, and right: the carbon chain-length distribution of the very-long-chain aliphatic (VLC) components and their cumulative contribution to the total VLC fraction (mean ± standard deviation, n = 5 biological replications).

The main cyclic components were lupeol and beta-amyrin with a cuticular wax coverage of 1.88 ± 0.50 µg cm^{-2} (23.0%) and 1.29 ± 0.18 µg cm^{-2} (15.8%), respectively (Table 18). *N*-alkanes were the most prominent VLC component class (24.4%, 1.99 ± 0.66 µg cm^{-2}), followed by primary alcohols (8.6%, 0.70 ± 0.11 µg cm^{-2}). Aditional VLC component classes detected in the cuticular wax were secondary alcohols, alkanoic acids, aldehydes, and alkyl esters. Carbon chain-lengths ranged from C_{20} to C_{52}, and the predominant chain-lengths were C_{29}, C_{30}, and C_{31} (Fig. 25). *N*-nonacosane dominated the VLC components (12.3%, 1.01 ± 0.36 µg cm^{-2}). The average carbon chain-length (ACL) of the VLC wax components was 31.84 carbon atoms.

Table 18. Cuticular wax composition of *Quercus coccifera* leaves from the Mediterranean woodland, southern Portugal (mean ± standard deviation, n = 5 biological replications).

Compound class	Carbon chain-length	Wax coverage (µg cm^{-2})
alkanoic acids	20	0.004 ± 0.002
	21	0.006 ± 0.003
	22	0.004 ± 0.001
	23	0.003 ± 0.001
	24	0.010 ± 0.005
	25	0.006 ± 0.003
	26	0.021 ± 0.006
	27	0.019 ± 0.006
	29	0.025 ± 0.010
	30	0.126 ± 0.045
	32	0.046 ± 0.024
primary alcohols	21	0.002 ± 0.001
	22	0.006 ± 0.004
	23	0.003 ± 0.001
	24	0.027 ± 0.008
	25	0.009 ± 0.002
	26	0.091 ± 0.027
	27	0.025 ± 0.003
	28	0.035 ± 0.012
	29	0.057 ± 0.015
	30	0.166 ± 0.025
	31	0.054 ± 0.005
	32	0.157 ± 0.039
	33	0.033 ± 0.006
	34	0.038 ± 0.010
alkyl esters	40	0.028 ± 0.006
	42	0.057 ± 0.008
	43	0.012 ± 0.002
	44	0.103 ± 0.004
	45	0.017 ± 0.002
	46	0.114 ± 0.005

Compound class	Carbon chain-length	Wax coverage ($\mu g\ cm^{-2}$)		
	47	0.020	±	0.004
	48	0.067	±	0.006
	49	0.012	±	0.003
	50	0.050	±	0.003
	51	0.015	±	0.003
	52	0.050	±	0.007
	28	0.007	±	0.003
aldehydes	30	0.052	±	0.024
	32	0.031	±	0.011
	25	0.048	±	0.038
	26	0.020	±	0.008
	27	0.272	±	0.139
	28	0.100	±	0.035
	29	1.005	±	0.361
n-alkanes	30	0.099	±	0.033
	31	0.361	±	0.067
	32	0.029	±	0.011
	34	0.029	±	0.008
	36	0.025	±	0.007
secondary alcohol	30	0.032	±	0.006
total very-long-chain aliphatics		*3.626*	*±*	*0.747*
beta-amyrin		1.288	±	0.293
beta-sitosterol		0.099	±	0.036
betulin		0.069	±	0.029
betulinic acid		0.062	±	0.033
erythrodiol		0.035	±	0.014
lupeol		1.876	±	0.503
oleanolic acid		0.063	±	0.017
triterpenoid 1		0.120	±	0.030
triterpenoid 2		0.050	±	0.027
total cyclic aliphatics		*3.663*	*±*	*0.787*
beta-tocopherol		0.050	±	0.023
delta-tocopherol		0.022	±	0.017
total cyclic aromatics		*0.071*	*±*	*0.036*
total cyclics		*3.734*	*±*	*0.811*
not identified		0.800	±	0.085
total wax		*8.160*	*±*	*1.568*

3.2.16 Cuticular leaf wax of Senna italica (hot desert)

Cuticular leaf wax of *S. italica* was qualitatively and quantitatively analysed. More than 94% of

the wax extract was identified using GC-MS. Total wax coverage of fully expanded leaflets of

S. italica amounted to 8.41 ± 0.72 µg cm^{-2} (mean ± standard deviation, n = 5 biological replicates). The leaf wax comprised a major fraction of very-long-chain acyclic (VLC) component classes (91.6% of the total wax, and wax coverage of 7.70 ± 0.77 µg cm^{-2}) and a minor fraction of cyclic components (2.7%, 0.22 ± 0.02 µg cm^{-2}).

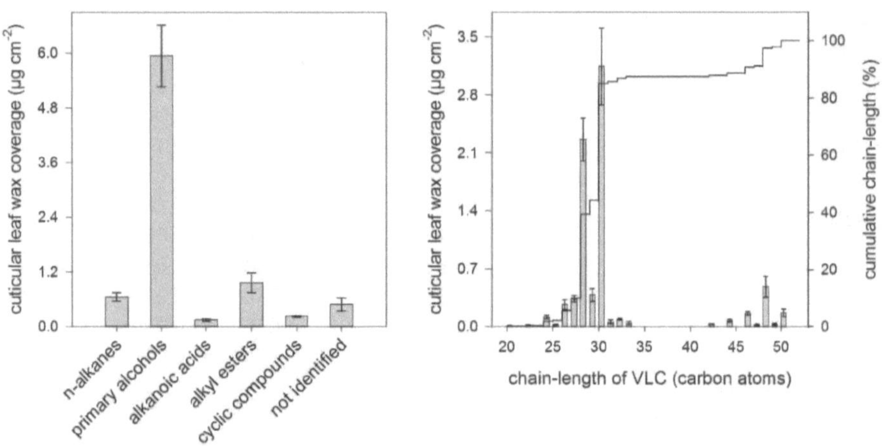

Fig. 26. Cuticular wax composition of *Senna italica* leaflets original from the hot desert, northeastern Saudi Arabia. Left: cuticular wax coverage arranged according to the component classes, and right: the carbon chain-length distribution of the very-long-chain aliphatic (VLC) components and their cumulative contribution to the total VLC fraction (mean ± standard deviation, n = 5 biological replications).

Primary alcohols were the most prominent VLC component class (70.7%, 5.94 ± 0.67 µg cm^{-2}), followed by alkyl esters (11.5%, 0.96 ± 0.22 µg cm^{-2}). Aditional VLC component classes detected in the cuticular wax were n-alkanes and alkanoic acids. Carbon chain-lengths ranged from C$_{20}$ to C$_{50}$, and the predominant chain-lengths were C$_{28}$, C$_{30}$, and C$_{48}$ (Fig. 26). Triacontanol dominated the VLC components (37.4%, 3.14 ± 0.46 µg cm^{-2}). The average carbon chain-length (ACL) of the VLC wax components was 31.20 carbon atoms. The main cyclic components were beta-amyrin and delta-amyrin with a cuticular wax coverage of 0.09 ± 0.01 µg cm^{-2} (1.1%) and 0.08 ± 0.02 µg cm^{-2} (0.9%), respectively (Table 19).

Table 19. Cuticular wax composition of *Senna italica* leaflets original from the hot desert, northeastern Saudi Arabia (mean ± standard deviation, n = 5 biological replications).

Compound class	Carbon chain-length	Wax coverage (µg cm^{-2})
alkanoic acids	20	0.010 ± 0.003
	22	0.008 ± 0.001
	24	0.102 ± 0.020
	26	0.025 ± 0.005
primary alcohols	22	0.007 ± 0.004
	24	0.013 ± 0.004
	26	0.231 ± 0.057
	27	0.050 ± 0.012
	28	2.263 ± 0.260
	29	0.147 ± 0.048
	30	3.143 ± 0.462
	32	0.087 ± 0.008
alkyl esters	42	0.028 ± 0.005
	44	0.072 ± 0.014
	46	0.163 ± 0.021
	47	0.023 ± 0.007
	48	0.483 ± 0.124
	49	0.028 ± 0.011
	50	0.167 ± 0.045
n-alkanes	25	0.023 ± 0.003
	26	0.009 ± 0.001
	27	0.285 ± 0.039
	29	0.233 ± 0.057
	31	0.056 ± 0.021
	33	0.042 ± 0.020
total very-long-chain aliphatics		*7.698 0.765*
alpha-amyrin		0.031 ± 0.017
beta-amyrin		0.093 ± 0.014
beta-sitosterol		0.012 ± 0.001
cholesterol		0.011 ± 0.004
delta-amyrin		0.076 ± 0.016
total cyclic aliphatics		*0.223 ± 0.016*
total cyclics		*0.223 ± 0.016*
not identified		0.485 ± 0.138
total wax		*8.405 ± 0.720*

3.2.17 Cuticular leaf wax of Smilax aspera (Mediterranean woodland)

Cuticular leaf wax of *S. aspera* was qualitatively and quantitatively analysed. More than 94% of the wax extract was identified using GC-MS. Total wax coverage of fully expanded leaves of *S. aspera* amounted to 4.63 ± 1.67 µg cm^{-2} (mean ± standard deviation, n = 5 biological replicates). The leaf wax comprised a major fraction of very-long-chain acyclic (VLC) component classes (94.6% of the total wax, and wax coverage of 4.89 ± 0.66 µg cm^{-2}) and cyclic components were absent.

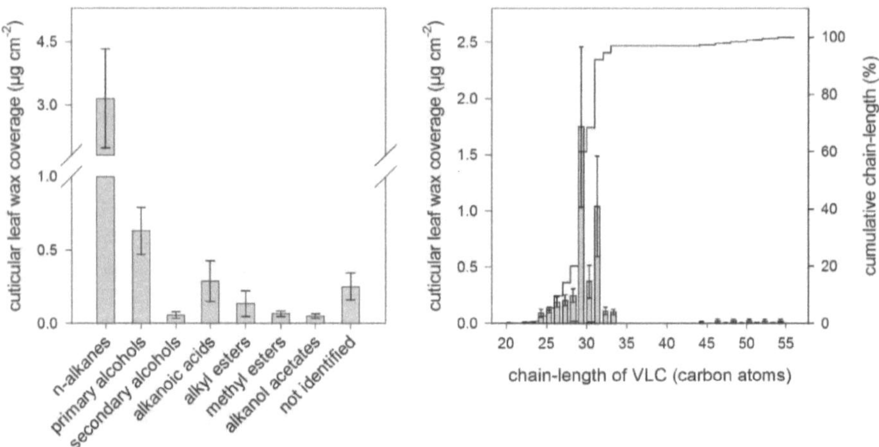

Fig. 27. Cuticular wax composition of *Smilax aspera* leaves from the Mediterranean woodland, southern Portugal. Left: cuticular wax coverage arranged according to the component classes, and right: the carbon chain-length distribution of the very-long-chain aliphatic (VLC) components and their cumulative contribution to the total VLC fraction (mean ± standard deviation, n = 5 biological replications).

N-alkanes were the most prominent VLC component class (68.2%, 3.15 ± 1.18 µg cm^{-2}), followed by primary alcohols (13.64%, 0.63 ± 0.16 µg cm^{-2}). Aditional VLC component classes detected in the cuticular wax were secondary alcohols, alkanoic acids, alkyl esters, methyl esters, and alkanol acetates. Carbon chain-lengths ranged from C_{20} to C_{54}, and the predominant chain-lengths were C_{29}, C_{30}, and C_{31} (Fig. 27). N-nonacosane dominated the VLC components (35.6%, 1.65 ± 0.69 µg cm^{-2}, Table 20). The average carbon chain-length (ACL) of the VLC wax components was 29.81 carbon atoms.

Table 20. Cuticular wax composition of *Smilax aspera* leaves from the Mediterranean woodland, southern Portugal (mean ± standard deviation, n = 5 biological replications).

Compound class	Carbon chain-length	Wax coverage (µg cm^{-2})
alkanoic acids	20	0.004 ± 0.001
	21	0.001 ± 0.000
	22	0.007 ± 0.001
	23	0.004 ± 0.001
	24	0.054 ± 0.021
	25	0.009 ± 0.003
	26	0.070 ± 0.032
	27	0.016 ± 0.007
	28	0.053 ± 0.035
	29	0.029 ± 0.017
	30	0.035 ± 0.024
	32	0.005 ± 0.003
primary alcohols	21	0.001 ± 0.000
	22	0.003 ± 0.000
	23	0.004 ± 0.001
	24	0.014 ± 0.003
	25	0.078 ± 0.022
	26	0.055 ± 0.010
	27	0.020 ± 0.005
	28	0.140 ± 0.030
	29	0.040 ± 0.010
	30	0.194 ± 0.075
	31	0.019 ± 0.006
	32	0.056 ± 0.019
	33	0.008 ± 0.003
alkyl esters	42	0.003 ± 0.000
	43	0.001 ± 0.000
	44	0.011 ± 0.006
	45	0.002 ± 0.000
	46	0.022 ± 0.016
	47	0.003 ± 0.001
	48	0.023 ± 0.015
	49	0.003 ± 0.001

Compound class	Carbon chain-length	Wax coverage ($\mu g\ cm^{-2}$)
	50	0.020 ± 0.014
	51	0.004 ± 0.001
	52	0.018 ± 0.015
	53	0.005 ± 0.001
	54	0.019 ± 0.014
	24	0.022 ± 0.011
methyl esters	26	0.028 ± 0.007
	28	0.017 ± 0.003
	25	0.033 ± 0.015
	26	0.038 ± 0.007
	27	0.158 ± 0.032
	28	0.038 ± 0.019
n-alkanes	29	1.646 ± 0.685
	30	0.092 ± 0.036
	31	1.007 ± 0.441
	32	0.049 ± 0.014
	33	0.093 ± 0.021
alkanol acetates	30	0.050 ± 0.015
	27	0.010 ± 0.006
secondary alcohols	29	0.031 ± 0.013
	31	0.014 ± 0.005
total very-long-chain aliphatics		*4.378 ± 1.579*
not identified		0.251 ± 0.090
total wax		*4.628 ± 1.666*

3.2.18 Cuticular leaf wax of Smilax japicanga (tropical dry forest)

Cuticular leaf wax of *S. japicanga* was qualitatively and quantitatively analysed. More than 94.5% of the wax extract was identified using GC-MS. Total wax coverage of fully expanded leaves of *S. japicanga* amounted to 5.18 ± 0.65 $\mu g\ cm^{-2}$ (mean ± standard deviation, n = 5 biological replicates). The leaf wax comprised a major fraction of very-long-chain acyclic (VLC) component classes (94.4% of the total wax, and wax coverage of 4.89 ± 0.66 $\mu g\ cm^{-2}$) and a minor fraction of cyclic components (0.5%, 0.03 ± 0.01 $\mu g\ cm^{-2}$).

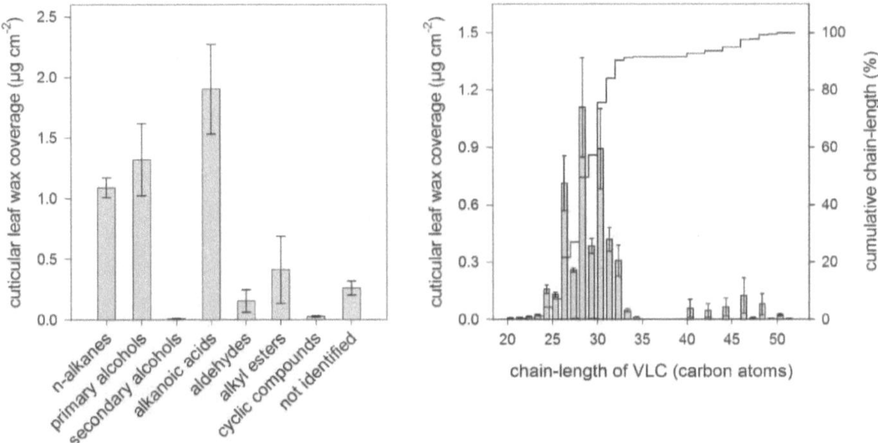

Fig. 28. Cuticular wax composition of *Smilax japicanga* leaves from the tropical dry forest, northeastern Brazil. Left: cuticular wax coverage arranged according to the component classes, and right: the carbon chain-length distribution of the very-long-chain aliphatic (VLC) components and their cumulative contribution to the total VLC fraction (mean ± standard deviation, n = 5 biological replications).

Alkanoic acids were the most prominent VLC component class (36.7%, 1.90 ± 0.37 µg cm^{-2}), followed by primary alcohols (25.5%, 1.32 ± 0.30 µg cm^{-2}). Aditional VLC component classes detected in the cuticular wax were n-alkanes, secondary alcohols, aldehydes and alkyl esters. Carbon chain-lengths ranged from C_{20} to C_{51}, and the predominant chain-lengths were C_{26}, C_{28}, and C_{30} (Fig. 28). Octacosanoic acid dominated the VLC components (11.4%, 0.59 ± 0.14 µg cm^{-2}). The average carbon chain-length (ACL) of the VLC wax components was 29.83 carbon atoms. The only two cyclic components were bis-(octylphenyl)-amine and cholesterol with a cuticular wax coverage of 0.02 ± 0.01 µg cm^{-2} (0.42%) and 0.005 ± 0.001 µg cm^{-2} (0.10%), respectively (Table 21).

Table 21. Cuticular wax composition of *Smilax japicanga* leaves from the tropical dry forest, northeastern Brazil (mean ± standard deviation, n = 5 biological replications).

Compound class	Carbon chain-length	Wax coverage ($\mu g\ cm^{-2}$)
alkanoic acids	20	0.007 ± 0.001
	21	0.003 ± 0.001
	22	0.010 ± 0.002
	23	0.014 ± 0.003
	24	0.141 ± 0.023
	25	0.030 ± 0.005
	26	0.461 ± 0.096
	27	0.058 ± 0.006
	28	0.592 ± 0.144
	29	0.063 ± 0.009
	30	0.398 ± 0.086
	31	0.024 ± 0.003
	32	0.099 ± 0.030
primary alcohols	21	0.008 ± 0.001
	22	0.005 ± 0.001
	23	0.009 ± 0.001
	24	0.016 ± 0.004
	25	0.026 ± 0.002
	26	0.169 ± 0.043
	27	0.024 ± 0.003
	28	0.417 ± 0.109
	29	0.031 ± 0.003
	30	0.403 ± 0.099
	31	0.024 ± 0.004
	32	0.179 ± 0.043
	34	0.011 ± 0.001
alkyl-esters	40	0.057 ± 0.047
	42	0.047 ± 0.035
	44	0.065 ± 0.047
	46	0.126 ± 0.092
	47	0.007 ± 0.004
	48	0.079 ± 0.055
	49	0.003 ± 0.001
	50	0.023 ± 0.006
	51	0.005 ± 0.001
aldehydes	26	0.023 ± 0.011
	27	0.015 ± 0.007
	28	0.035 ± 0.021
	29	0.014 ± 0.007
	30	0.039 ± 0.025
	32	0.029 ± 0.021
n-alkanes	25	0.073 ± 0.012
	26	0.053 ± 0.009
	27	0.159 ± 0.022

Compound class	Carbon chain-length	Wax coverage ($\mu g\ cm^{-2}$)
	28	0.061 ± 0.006
	29	0.273 ± 0.030
	30	0.053 ± 0.021
	31	0.369 ± 0.057
	33	0.047 ± 0.009
secondary alcohols	26	0.005 ± 0.001
	28	0.004 ± 0.001
total very-long-chain aliphatics		*4.890 ± 0.657*
cholesterol		0.005 ± 0.001
total cyclic aliphatics		*0.005 ± 0.001*
bis-(octylphenyl)-amine		0.022 ± 0.006
total cyclic aromatics		*0.022 ± 0.006*
total cyclics		*0.030 ± 0.010*
not identified		0.261 ± 0.058
Total wax		*5.178 ± 0.646*

3.3 Discussion

Cuticular permeability varies among plant species from 2.55 x 10^{-7} m s^{-1} (*Zamioculcas zamiifolia,* adapted to arid subtropical areas) to 4.00 x 10^{-3} m s^{-1} (*Ipoema batatas,* widely grown in tropical, subtropical, and warm temperate regions; Schuster *et al.,* 2017). Earlier attempts to categorise cuticular permeances according to the plant original habitat showed that tropical evergreen epiphytic or climbing plants possessed the lowest permeance values, followed by Mediterranean evergreen, and temperate deciduous plants had the highest permeances (Schreiber and Riederer, 1996; Riederer and Schreiber, 2001). Also, at a congeneric level, the cuticular conductances of 11 *Quercus* species growing in a common garden were investigated, and the g_{min} of Mediterranean evergreen oaks slightly differed from that of temperate deciduous oaks (Gil-Pelegrín *et al.,* 2017). Therefore, one might expect that plants from arid environments have lower permeabilities in comparison with plants from humid environments.

However, our findings showed that xerophilous plants do not possess unusually low cuticular water permeability, as well as an adaptation of cuticular permeability to original habitat was not identified (Fig. 14). In line with our results, Schuster *et al.* (2017) reported that the median of leaf cuticular permeability of 160 plant species from widely different habitats was 3.60 x 10^{-5} m s^{-1}, and the central 50% of the species comprised cuticular permeabilities ranging from 1.02 x 10^{-5} to 7.20^{-5} m s^{-1}. In our survey of cuticular transpiration barrier, except for the two *Smilax* species (climbing plants), the g_{min} varies among plant species from 1.11 ± 0.24 x 10^{-5} m s^{-1} (*P. dactylifera*, a desert plant) to 6.94 ± 2.00 x 10^{-5} m s^{-1} (*C. colocynthis*, also a desert plant), which lies between the central 50% of cuticular permeability described above. The low g_{min} values found for *S. japicanga,* and *S. aspera* (0.46 ± 0.24 and 0.83 ± 0.15 x 10^{-5} m s^{-1}, respectively) properly fit within the values so far found for evergreen vines and lianas (Schuster *et al.,* 2017).

Spatial variability in abiotic filters can drive different plant life strategies, which might explain the lack of correlation between cuticular transpiration barrier and habitat. A strictly controlled experiment was conducted with genetically related plants of *Quercus coccifera* growing under

Mediterranean arid conditions and temperate coastal conditions to identify potential influence of the growth environment on the cuticular transpiration barrier, and the efficiency of the cuticular transpiration barrier was unresponsive to environmental features (see chapter one for details). We provided substantial evidence for adoption of alternative plant strategies to cope with the same stressors within a shared environment. The two desert plants *C. colocynthis* and *P. dactylifera*, as shown in chapter two, represent a classical and illustrative example of different life strategies to deal with water and heat stresses, which culminate in distinct efficiencies of their cuticular transpiration barrier. In short, the g_{min} of *P. dactylifera* at mild temperature was about 6-fold lower than that of *C. colocynthis*. Also, the increase of temperature from 25 to 50°C had a minor effect on the cuticular transpiration barrier of *P. dactylifera*, whereas drastically compromised the function of the barrier against water loss of *C. colocynthis*.

In our study, we also surveyed T_{crit} values of flowering plant species from some of the hottest and driest sites in South-America, Europe, and Asia. P_{TT} values of the studied species varied by more than 5°C and it is in line with previous studies on desert plants (Curtis *et al.*, 2014, 2016). Although plants displayed high plasticity in photosynthetic thermal tolerance (P_{TT}), the selected temperature threshold (T_{crit}) did not differ in response to the species original habitat (Fig. 15). Teskey *et al.* (2015) reported that the severity of heat damage to leaf physiological functions increases in periods of reduced water availability. However, we also found evidence in the literature that drought can increase the leaf photosynthetic thermal tolerance (P_{TT}) at the species-specific level. In some cases, exceeding the effects of elevated growth temperatures (Ladja *et al.*, 2000; Ghouil *et al.*, 2003). Within an arid climatic domain, Curtis *et al.* (2016) investigated the P_{TT} of 42 plant species and stressed that plants adapted to wetter microhabitat are more susceptible to thermal damage than those plants adapted to microhabitats that are more arid. Thereby suggesting a drought-mediated heat tolerance, which is in line with our Hypothesis that plants better adapted to drought (those with more efficient cuticular transpiration barrier) possess higher leaf photosynthetic thermal tolerance.

Our findings showed a strongly significant correlation between cuticular transpiration barrier and leaf thermal tolerance (Fig. 16). Plants with more efficient cuticular transpiration barriers had higher thermal tolerance. Thereby suggesting a drought-mediated heat tolerance. However, a drought-mediated heat tolerance might be further investigated given the contradictory nature of previous studies. Knight and Ackerly (2003), using T_{crit} as a proxy of P_{TT}, compared four pairs of desert and coastal congeneric species and found no difference in P_{TT}. By contrast, Zhu *et al.* (2018) reported in a study on 62 plant species native to five thermally distinct biomes that plant species adapted to hot biomes possess inherently higher P_{TT} (quantified as T_{crit}) than those adapted to mild biomes, thereby suggesting that T_{crit} is highly temperature dependent. Nevertheless, variations in P_{TT} were not linked to precipitation level at the species original habitat.

In conclusion, for the first time, the efficiency of cuticular transpiration barrier (quantified as g_{min}) and the leaf photosynthetic thermal tolerance (quantified as T_{crit}) of 14 flowering plant species native to three hot arid biomes in three continents were concomitantly investigated. We found high variability for cuticular transpirational barrier and leaf thermal tolerance among species, but both physiological features were unresponsive to the species original habitat. Also, we provide substantial evidence that xerophilous flowering plants with more efficient cuticular transpiration barrier possess higher leaf thermal tolerance, which might indicate potential convergent evolution of these two plant features in hot arid biomes. Therefore, we suggest that the high variability of both cuticular transpiration barrier and thermal tolerance in arid biomes might reflect the different life strategies adopted by plants to cope with drought and heat driven by abiotic filters.

General discussion

Living in arid environments is challenging for all organisms, particularly for plants that cannot escape adverse environmental conditions through movement or behaviour. Plant evolution has resulted in uncountable properties and strategies that allow plants to cope with the high dehydrating terrestrial environment and its associated variation of temperatures, thereby permiting plants to colonise from mild to extremely harsh climate domains across the globe. Plant cuticle is a synapomorphic feature of higher plants, and it represents one of the most critical adaptative traits for plant survival in the land (Yeats and Rose, 2013). Here, for the first time, we concomitantly investigated the cuticular transpirational barrier and the leaf photosynthetic thermal tolerance of 14 non-succulent plants species native from three hot arid biomes distributed in three continents (Brazilian dry forest, Portuguese Mediterranean woodland, and Saudi desert). This study comprised several phylogenetically independent flowering species including a congeneric *Smilax* species pair, which differed in their natural habitat (*S. japicanga* grows in the Brazilian dry forest, whereas *S. aspera* occurs in the Portuguese Mediterranean woodland).

The cuticular permeability of non-succulent xerophytes

Drought-stressed plants close the stomata to minimise the transpirational water loss. Under this conditions, the remained water loss only can occur across the cuticle. Thus, the cuticular permeability corresponds to the minimal leaf water loss when its stomata are maximally closed (Körner, 1995). The minimum leaf conductance (g_{min}) was used as a proxy for cuticular water permeability. Cuticular permeabilities of the studied xerophilous plants ranged from $0.46 \pm 0.24 \times 10^{-5}$ m s^{-1} (*Smilax japicanga*, a tropical dry forest climbing plant) to $6.94 \pm 2.00 \times 10^{-5}$ m s^{-1} (*Citrullus colocynthis*, a hot desert plant). Cuticular permeability varies among plant species from 2.55×10^{-7} m s^{-1} (*Zamioculcas zamiifolia*, adapted to arid subtropical areas) to 4.00×10^{-3} m s^{-1} (*Ipoema batatas*, widely grown in tropical, subtropical, and warm temperate regions) (Schuster *et al.*, 2017; Karbulková *et al.*, 2008; Zobayed *et al.*, 2000).

Previous attempts to categorise cuticular permeances according to the plant original habitat showed that tropical evergreen epiphytic or climbing plants possessed the lowest permeance values, followed by Mediterranean evergreen plants, and temperate deciduous plants had the highest permeances (Schreiber and Riederer, 1996; Riederer and Schreiber, 2001). Also, at a congeneric level, the cuticular conductances of 11 *Quercus* species growing in a common garden were investigated, and the g_{min} of Mediterranean evergreen oaks were slightly lower than that found for temperate deciduous oaks (Gil-Pelegrín *et al.*, 2017). Therefore, one might intuitively expect that plants native from arid regions have lower permeabilities than those plants from humid climates.

A recent study reviewed the last 20-years research on cuticular permeability and reported that the median of leaf cuticular permeability of 160 plant species from widely different habitats was 3.60×10^{-5} m s^{-1}. The central 50% of the species comprised cuticular permeabilities ranging from 1.02×10^{-5} m s^{-1} to 7.20×10^{-5} m s^{-1} (Schuster *et al.*, 2017). In our survey of permeabilities, except for the two *Smilax* species, the g_{min} ranged from $1.11 \pm 0.24 \times 10^{-5}$ m s^{-1} (*Phoenix dactylifera*, a desert plant) to $6.94 \pm 2.00 \times 10^{-5}$ m s^{-1} (*Citrullus colocynthis*, also a desert plant). Therefore, the g_{min} of 12 out of 14 non-succulent plant species lie between the central 50% of cuticular permeabilities described above. *Smilax japicanga* and *Smilax aspera* had the lowest permeabilities (0.46 ± 0.24 and $0.83 \pm 0.15 \times 10^{-5}$ m s^{-1}, respectively) among the 14 plant species. The two *Smilax* species are evergreen climbing plants and their permeabilities adequately fit within the permeabilities so far reported for evergreen vines and lianas, such as *Hedera helix* and *Vanilla planifolia* (Schuster *et al.*, 2017). Therefore, our findings do not support the usual assertion that plants from arid regions have notably lower permeabilities in comparison with plants from humid environments. In line with these results, Schuster *et al.* (2016) reported that the desert evergreen *Rhazya stricta* does not possess a specially low cuticular permeability (5.4×10^{-5} m s^{-1}). However, this was the first report in the literature, wherein a plant species was capable of maintaining the effectiveness of the cuticular transpiration barrier at high temperatures.

The plant cuticle as leaf cuticular transpiration barrier in hot arid biomes

The plant cuticle covers the epidermal cells of all land plants and consists of the insoluble fatty acid polyester cutin and waxes embedded within and deposited on top of the cutin framework (Jeffree, 2006). The cuticular waxes constitute the major barrier against water diffusion from the leaf interior to the atmosphere (Schönherr, 1976). The total cuticular wax yield varies among plant species and responds to biotic and abiotic stresses. The extraction of the cuticular waxes leads to an increase of the cuticular water permeability by up to several orders of magnitude (Schönherr, 1976; Schönherr and Lendzian, 1981).

The leaf cuticular wax chemistry of 14 plant species was analysed qualitatively and quantitatively. The corresponding minimum leaf conductance was determined (Table 14). The plants displayed a wide range of leaf wax content. The desert plant *Citrullus colocynthis* had the lowest total wax (4.20 ± 0.44 µg cm^{-2}) and the Mediterranean *Phillyrea angustifolia* exhibited the highest wax amount (310.70 ± 3.84 µg cm^{-2}). Cyclic wax components were absent in the wax mixtures of the Mediterranean plants *Chamaerops humilis* and *Smilax aspera* and varied from 0.008 ± 0.003 µg cm^{-2} (the desert plant *Citrullus colocynthis*) to 277.15 ± 5.26 µg cm^{-2} (the Mediterranean plant *Phillyrea angustifolia*). Very-long-chain aliphatic (VLC) fraction varied from 3.63 ± 0.75 µg cm^{-2} (the Mediterranean plant *Quercus coccifera*) to 25.66 ± 3.65 µg cm^{-2} (the desert plant *Phoenix dactylifera*). Weighted average carbon-chain-length (ACL) ranged from 29.81 (the Mediterranean plant *Smilax aspera*) to 41.02 (the desert *Phoenix dactylifera*).

The potential relationship between minimum leaf conductance and cuticular wax chemistry was further investigated using correlation analyses. For this purpose, we excluded the congeneric *Smilax* species pair from the investigations. There is substantial evidence in the literature that evergreen vines and lianas constitute a group with notably lower cuticular permeabilities (Riederer and Schreiber, 2001; Schuster *et al.*, 2017) and it is likely that these plants have differentiated mechanisms to avoid non-stomatal water loss.

The g_{min} of 12 xerophilous plant species did neither correlate with the total wax coverage nor with the cyclic wax coverage. However, the natural logarithm of the g_{min} was negatively correlated with the VLC aliphatic fraction and ACL (Fig. 29).

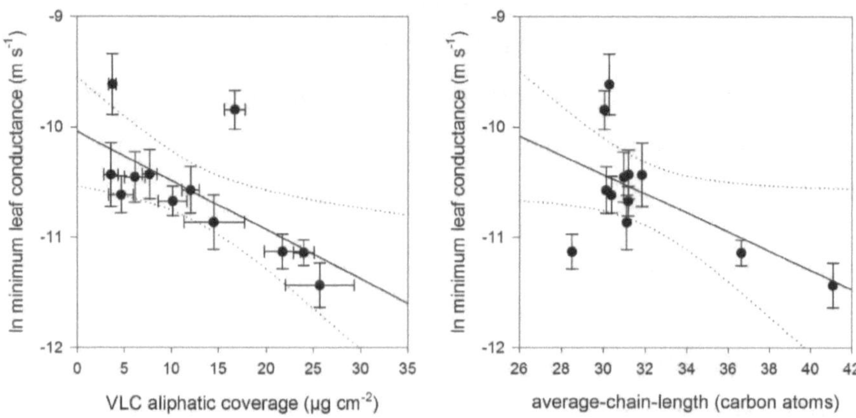

Fig. 29. Natural logarithm of the minimum leaf conductance of 12 xerophilous non-succulent plants as a function of the total very-long-chain aliphatic coverage (left, R^2 = 0.47, p = 0.02) and the weighted average-chain-length (right, R^2 = 0.34; p = 0.05). Dashed lines represent 95% confidence intervals.

These findings are in agreement with the potential functions of very-long-chain aliphatic compounds. Previous experiments with tomatoes fruits (Vogg *et al.*, 2004; Leide *et al.*, 2007, 2011), *Arabidopsis* leaves (Buschhaus and Jetter, 2012) and leaves of eight evergreen woody plant species (Jetter and Riederer, 2016), suggest that the VCL aliphatics majorly make up the cuticular transpiration barrier. Moreover, a recent study investigated 17 plant species native from widely different habitats and found that the VLC aliphatic wax fraction constituted circa 55% of the cuticular transpiration barrier (Schuster, 2016). Also, Huang (2017) investigated the cuticular transpiration barrier of fruits and leaves of 17 crop plant species, and the results showed that, in general, leaves had lower cuticular permeability than that found to corresponding fruits. The author suggests that the longer ACLs of leaves constitute a more effective transpiration barrier than the shorter ACLs found for fruits. Therefore, to further investigate the cuticular transpiration barrier of xerophilous plants, the g_{min} was predicted from the VLC aliphatic wax fraction and its corresponding ACL (Fig. 30). The combination of the VLC aliphatic coverage and ACL was suitable to predict the minimum conductance with a slope of 0.52. The equation *ln g_{min} = -8.825 - (0.0340 * VLC aliphatics) - (0.0422 * ACL)* was obtained from the multiple linear regression of the natural logarithm of the g_{min} as a function of VLC aliphatic coverage and ACL.

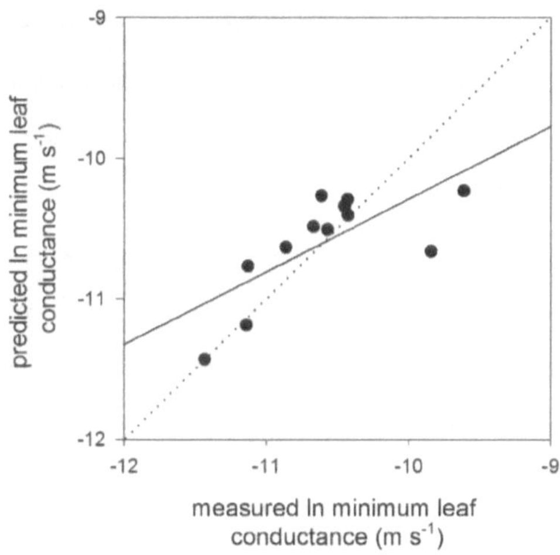

Fig. 30. Natural logarithm of the minimum leaf conductance of 12 xerophilous non-succulent plants from an equation based on the very-long-chain (VLC) aliphatic wax coverage and average-chain-length of VLC aliphatic components. Plot of predicted versus measured ln minimum leaf conductance (R^2 = 0.52; p = 0.01). The dashed line represents a slope equal to one.

The studied plants showed high heterogeneity of the chemical composition of the cuticular waxes. VLC aliphatic component classes, for example, *n*-alkanes, alkanoic acids, primary alcohols, and alkyl esters were common constituents fo the cuticular waxes. Also, cyclic components were abundant in the cuticular waxes of many plant species. In line with previous studies, the minimum leaf conductance of the studied plants did not decrease with increasing total leaf wax coverage. Thus, higher leaf cuticular wax amounts do not constitute more effective cuticular transpiration barrier (Schreiber and Riederer, 1996; Riederer and Schreiber, 2001; Schuster, 2016). Cyclic compounds play a role in the stabilisation of heat-stressed plant cuticle (Schuster *et al.*, 2016), but its contribution to avoid the cuticular water loss is minimal or absent (Vogg *et al.*, 2004; Leide *et al.*, 2007, 2011; Buschhaus and Jetter, 2012). These findings corroborate the model proposed for the molecular structure of cuticular waxes by Riederer and Schreiber (1995). The authors suggested that cuticular waxes are multiphase systems composed of highly ordered crystalline domains interspersed with amorphous zones. According to this model, the inner parts of the alkyl chains of the VLC aliphatic molecules make up the crystalline domains of the cuticular waxes, while the irregularly protruding chain ends and other components, which do not fit into the crystallites, form the adjacent amorphous zones. This model predicts that the VLC fraction of the cuticular waxes mainly represents the cuticular transpirational barrier in plants. The average-chain-length of VLC aliphatic components are believed to enhance crystalline domains and thereby contributing to the efficiency of the cuticular transpiration barrier (Riederer and Schneider, 1990; Riederer, 1991).

The effect of temperature on the cuticular transpiration barrier of non-succulent xerophytes

Temperature is a critical ecological factor in arid environments. Even under temperate climate conditions, plant leaves may reach temperature as high as 40°C. The temperature effect on the cuticular water permeability decisively influences the physiological and ecological role of the plant cuticle, especially as high evaporative demand and elevated temperatures very often occur together. Therefore, the thermal stability of the cuticular transpirational barrier is decisive for safeguarding xerophilous plants against desiccation.

The diffusion of water in the cuticular transpiration barrier is a kinetic phenomenon and, consequently, rises with temperature. In all cases studied so far, high temperatures also led to an increased cuticular permeability (Riederer, 2006; Schreiber, 2002). The temperature-dependence of a kinetic parameter can be analysed quantitatively and mechanistically using the Arrhenius formalism. For this purpose, one can plot the natural logarithm of the rate constant (in this case the cuticular permeability) against the inverse absolute temperature. An abrupt increase or decrease of the slope of the Arrhenius plot at a specific temperature indicates a change in the properties of the surroundings of the diffusing water molecule. The cuticular permeabilities of all non-desert plants investigated so far have slightly increased at temperatures from 15°C to 35°C, whereas temperatures above 35°C caused a drastic rise in the permeability (Schreiber, 2001; Riederer, 2006). To date, a single hot desert plant (*Rhazya stricta*) has been investigated in details. In contrast to the already studied non-desert plants, the cuticular water permeability of *R. stricta* marginally increased between 15°C and 50°C without an abrupt change in slope (Schuster *et al.*, 2016).

In this study, we further investigated the temperature effect on the cuticular transpiration barrier of the desert water-saver *Phoenix dactylifera* and the water-spender *Citrullus colocynthis*. The Arrhenius formalism applied to the cuticular permeances of *C. colocynthis* leaves led to two distinct linear plots for the lower and higher temperature range, respectively: (1) a slight increase of permeance at lower temperatures ranging from 25-40°C and (2) a sharp increase at higher temperatures from 40°C to 50°C (Fig. 11B). Moreover, the thermal-dependent increase of permeability of *C. colocynthis* raised by a factor of 3.2 over the whole experimental temperature range (25-50°C). These findings indicate that at higher temperatures the water molecules diffuse through pathways not accessible at lower temperatures, which require much higher activation energy to facilitate diffusive movement (Riederer, 2006).

In contrast to *C. colocynthis*, the Arrhenius plot for the water permeability of *P. dactylifera* is linear over the whole range of temperatures studied (Fig. 11B). Permeability increased slightly from 25°C to 50°C, and the permeability raised only by a factor of 1.2 in *P. dactylifera*. This pronounced resistance against a thermally induced breakdown of the cuticular transpiration barrier is comparable to that recently described for the hot desert shrub *R. stricta* (Schuster *et al.*, 2016) suggesting that resistance against pronounced thermal effects on cuticular permeability might be an adaptive trait of desert water-saver plants.

The physical and chemical basis for the thermal stability of the cuticular transpiration barrier of *R. stricta* was attributed to the composition and physical structure of its cuticle. The cuticular wax composition of *R. stricta* comprises a remarkably high proportion of pentacyclic triterpenoids (85% of the total wax; Schuster *et al.*, 2016). There is evidence that triterpenoids and cutin matrix are intermixed in the interior of the plant cuticle (Tsubaki *et al.*, 2013). In the case of *R. stricta*, the mass-based triterpenoid-to-cutin ratio was 0.63, which indicates that 63% of the cutin polymer is linked to agglomerations of triterpenoids (Schuster *et al.*, 2016). According to the authors, the triterpenoids fill voids within the cutin polymer, thereby fixating the polymer strands. The resulting intermix of triterpenoids and cutin matrix would increase the density and the mechanical strength of the cuticle as a whole. Hence, the maintenance of the cuticular transpiration barrier at high temperatures in *R. stricta* is achieved by the mechanical enhancement of the cuticular matrix, which reduces the thermal stress on the layer of VLC aliphatic cuticular waxes (Schuster *et al.*, 2016). Although the thermal stability of the cuticular transpiration barrier of *P. dactylifera* is comparable to that of *R. stricta*, this model depicting the localisation and function of the triterpenoid fillers cannot be applied to the former species because the cuticular wax composition of *P. dactylifera* is almost exclusively made up of VLC aliphatic compounds (88% of the total wax) and only a minor part of triterpenoids (4%).

So, the question is how does *P. dactylifera* achieve resistance against a thermally induced breakdown of the cuticular transpiration barrier? To shed some light on this question is necessary to address the physical and chemical basis for the difference found in the thermal susceptibility of the cuticular transpiration barrier of the water-saver *P. dactylifera* and the water-spender *C. colocynthis*. One can assume that the observed differences are related to the composition and physical structure of the cuticle of both species. Especially, those concerning the qualitative and quantitative composition of the main components of the cutin and waxes. The cutin matrix of both plant species mainly consisted of ester-linked C_{16} and C_{18} hydroxy alkanoic acids predominantly 9,16- and 10,16-dihydroxy hexadecanoic acids in *C. colocynthis* leaves and 18-hydroxy 9,10-epoxy octadecanoic acid in *P. dactylifera* leaflets. The cutin composition of *C. colocynthis* leaves represents a C_{16} type, and *P. dactylifera* leaflets a C_{18} type of cutin according to the classification proposed by Holloway (1982). The amount of cutin per unit surface area was seven-fold lower in *C. colocynthis* leaves as compared to *P. dactylifera* leaflets. The presence of 18-hydroxy 9,10-epoxy octadecanoic acid in cutins is correlated with the occurrence of non-ester cutin fractions additionally cross-linked by mid-chain ether bonds (Riederer and Schönherr, 1988). The resulting cutin polymer is highly recalcitrant (Schmidt and Schönherr, 1982) and may, therefore, reinforce the cuticle of *P. dactylifera* against thermal damage at high temperatures.

Differences in the composition of the cuticular waxes, specifically in their VLC aliphatic fraction, constitute another reason for the differences in the thermal susceptibility of the cuticular transpiration barrier. Although *P. dactylifera* and *C. colocynthis* qualitatively present similar aliphatic compound classes, the quantitative contribution of each class to the total cuticular wax load and the carbon chain-length of homologous compounds considerably vary between the two species. The cuticular waxes of *P. dactylifera* contain higher amounts of all identified compound classes, except for comparable amounts of primary alcohols. The total VLC aliphatic wax coverage of the leaflets of this species was seven-fold higher than in *C. colocynthis* leaves. Moreover, the patterns of the respective chain-length distributions of VLC aliphatic wax components drastically differ between the two species. The carbon chain-lengths of the VLC aliphatic compounds range from C_{20} to C_{62} in *P. dactylifera* and from C_{20} to C_{42} in *C. colocynthis* (Fig. 13). Therefore, the VLC aliphatics of both species can be divided into two groups: (1) components with less than 40 carbon atoms and (2) components with chains of 40 and more carbon atoms. Alkanes, alkenes, aldehydes, alkanoic acids and, primary and secondary alcohols make up the fraction with chain-lengths $< C_{40}$, whereas the wax fraction with chain-length $\geq C_{40}$ consists exclusively of alkyl esters. The aliphatic components with chain-lengths $< C_{40}$ and $\geq C_{40}$ comprise 69 and 31 mol %, respectively, of the total VLC aliphatic fraction in *P. dactylifera*. In contrast, only a minor fraction (0.65 mol %) of the VLC aliphatic components was $\geq C_{40}$ in *C. colocynthis* while almost all compounds belonged to the shorter chain-length fraction of the wax.

Thus, we hypothesise that the high proportion of VLC aliphatic compounds $\geq C_{40}$ in *P. dactylifera* significantly affects the physical properties of the leaf cuticular transpiration barrier, thereby leading to more efficient control of the cuticular water loss over the range of extreme temperatures experienced by this plant in its natural habitat.

The basis for this peculiar thermal stability may lie in the high melting range of plant waxes containing a considerable amount of VLC alkyl esters. Carnauba wax, a cuticular wax of the carnauba palm (*Copernicia prunifera*) with an alkyl ester content comparable to that of *P. dactylifera* has a melting range from 65° to 90°C (Basson and Reynhardt, 1988). Intuitively, one might expect that the reason for this high melting range is the formation of distinct crystalline domains consisting of alkyl esters with chain-lengths $\geq C_{40}$ with thermal properties differing from those of domains made up by the shorter chain wax components. This would be in line with findings from paraffine waxes wherein compounds associated in distinct crystalline domains if their chain-lengths differed by more than five carbon atoms (Retief and le Roux, 1983).

However, the segregation of the VLC alkyl esters from the other wax components would not lead to a high-melting crystalline domain. The melting points of the esters of VLC alcohols and VLC acids have melting points considerably lower than the corresponding VLC alkanoic acids, alcohols or alkanes (Fig. 31). The weighted median chain-length of the ester fraction of *P. dactylifera* wax is 51, and it has an estimated melting point of circa 87°C. Wax components with approximately the same melting point are the C_{26} alkanoic acid and the C_{31} alcohol. Hence, a compound can be synthesised with circa half the expenditure of energy and carbon, in the case of the acid, and thereby having the same melting point as the C_{51} ester. The base for the comparably low melting points of VLC compounds with an ester group in the middle part of the carbon chain is the disorder introduced by this group into the crystalline packing.

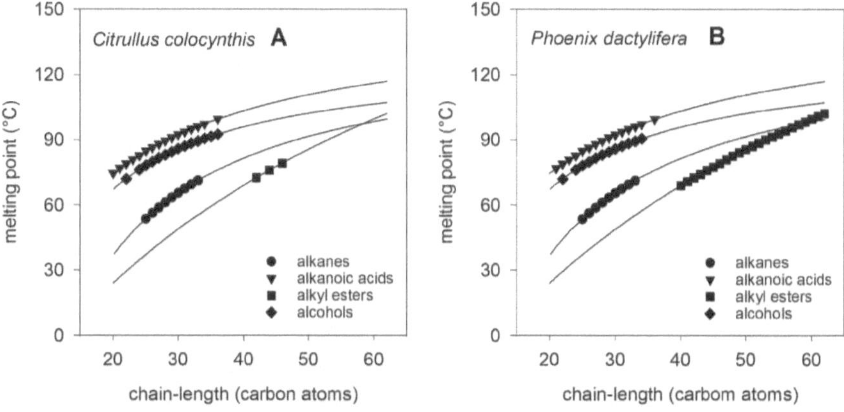

Fig. 31. Estimated melting points of single very-long-chain alkanes, alkanoic acids, alkyl esters, and alcohols present in the cuticular waxes from leaves of *Citrullus colocynthis* (A) and leaflets of *Phoenix dactylifera* (B). The melting points were obtained from curves fitted to experimental data (alkyl esters: Iyengar and Schlenk, 1969; all other compounds: Small, 1984).

The presence of VLC esters in the wax of *P. dactylifera* and other plant species, therefore, should have different reasons. It has been shown that VLC esters found in plant waxes do not form crystalline domains separate from those made up of the remaining long-chain wax components of roughly half their chain-length. Instead, as suggested earlier, electron diffraction studies showed that VLC compounds with a central ester group integrate within two adjacent crystalline lamellae composed of the other long-chain compounds (Dorset, 2005, 2002 and 1999; Reinhardt and Riederer, 1994). These bridging molecules form covalent anchors holding the crystallites together (Fig. 32B) and thereby increasing the melting point of the wax and, consequently, its thermal mechanical stability.

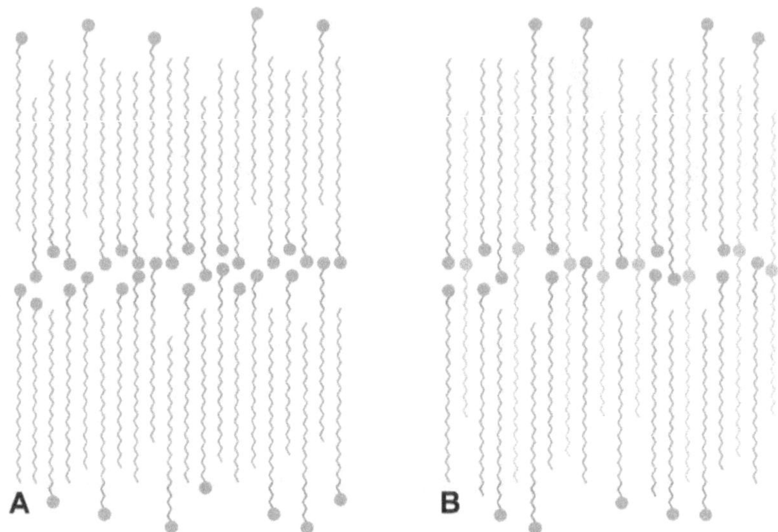

Fig. 32. Model proposed for the molecular structure of the cuticular waxes of *Citrullus colocynthis* leaves (A) and *Phoenix dactylifera* leaflets (B). The inner parts of the alkyl chains of very-long-chain (VLC) aliphatic molecules (blue zigzag lines) form the highly ordered crystalline domains (blue background), whereas the chain ends make up the adjacent amorphous zones of the cuticular waxes. Dots symbolise functional groups. The VLC esters (orange zigzag lines) integrate within two crystalline lamellae. These bridging molecules form covalent anchors holding the crystallites together, thereby increasing the mechanical stability of the *P. dactylifera* cuticular transpiration barrier at higher temperatures.

Last, we proposed that the small temperature effect on the cuticular transpiration barrier of *P. dactylifera* in comparison to *C. colocynthis* be due to the higher mechanical stability of the ester-containing waxes of the former species. The presence of ester-containing "bridged" waxes might mechanically reinforce the cuticular transpiration barrier of *P. dactylifera* as a whole and, thus, prevent the occurrence of thermally induced structural defects that lead to abrupt increases of water permeability at higher temperatures as observed for *C. colocynthis* and many other non-desert species (Riederer, 2006). The production of ester-reinforced cuticular waxes might be an adaptation to hot environments similar to that described for *Rhazya stricta* (Schuster *et al.*, 2016) but achieved by the synthesis of an appreciable amount of VLC alkyl esters instead of triterpenoids.

References

Althawadi AM, Grace J. 1986. Water use by the desert cucurbit *Citrullus colocynthis* (L.) Schrad. Oecologia **70**, 475-480.

Baquedano FJ, Castillo FJ. 2006. Comparative ecophysiological effects of drought on seedlings of the Mediterranean water-saver *Pinus halepensis* and water-spenders *Quercus coccifera* and *Quercus ilex*. Trees **20**, 689-700.

Barrow SC. 1998. A Monograph of *Phoenix* L. (Palmae: Coryphoideae). Kew Bulletin **53**, 513-575.

Basson I, Reynhardt EC. 1988. An investigation of the structures and molecular dynamics of natural waxes: II. Carnauba wax. Journal of Physics D: Applied Physics **21**, 1429-1433.

Bauer S, Schulte E, Thier HP. 2004. Composition of the surface wax from tomatoes. I. Identification of the components by GC/MS. European Food Research and Technology **219**, 223-228.

Beerling DJ, Osborne CP, Chaloner WG. 2001. Do drought-hardened plants suffer from fever? Trends in Plant Science **6**, 507-508.

Burghardt M, Riederer M. 2003. Ecophysiological relevance of cuticular transpiration of deciduous and evergreen plants in relation to stomatal closure and leaf water potential. Journal of Experimental Botany **54**, 1941-1949.

Burghardt M, Riederer M. 2006. Cuticular transpiration. In: Riederer M, Müller C, eds. Biology of the plant cuticle, Vol. 23. Oxford: Blackwell Publishing, 292-311.

Buschhaus C, Jetter R. 2012. Composition and physiological function of the wax layers coating *Arabidopsis* leaves: β-amyrin negatively affects the intracuticular water barrier. Plant Physiology **160**, 1120-1129.

Cameron KD, Teece MA, Smart LB. 2006. Increased accumulation of cuticular wax and expression of lipid transfer protein in response to periodic drying events in leaves of tree tobacco. Plant Physiology **140**, 176-183.

Castro-Díez P, Navarro J. 2007. Water relations of seedlings of three *Quercus* species: variations across and within species grown in contrasting light and water regimes. Tree Physiology **27**, 1011-1018.

Coley PD, Bryant JP, Chapin FS III. 1985. Resource availability and plant antiherbivore defense. Science **230,** 895-899.

Cunningham GM, Mulham WE, Milthorpe PL, Leigh JH. 2011. Plants of western New South Wales. Csiro Publishing, Collingwood, Australia.

Curtis EM, Gollan J, Murray BR, Leigh A. 2016. Native microhabitats better predict tolerance to warming than latitudinal macro-climatic variables in arid-zone plants. Journal of Biogeography **43,** 1156-1165.

Curtis EM, Knight CA, Petrou K, Leigh A. 2014. A comparative analysis of photosynthetic recovery from thermal stress: a desert plant case study. Oecologia **175,** 1051-1061.

De Micco V, Aronne G. 2012. Morpho-anatomical traits for plant adaptation to drought. In: Aroca R, ed. Plant responses to drought stress. Berlin: Springer, 37-61.

Díaz S, Kattge J, Cornelissen JH *et al.* 2016. The global spectrum of plant form and function. Nature **529,** 167-171.

Dorset DL. 1999. Bridged lamellae: crystal structure(s) of low molecular weight linear polyethylene. Marcromolecules **32,** 162-166.

Dorset DL. 2002. From waxes to polymers – crystallography of polydisperse chain assemblies. Structural chemistry **13,** 329-337.

Dorset DL. 2005. Crystallography of the polymethylene chain: an inquiry into the structure of waxes. Oxford University Press on Demand, England.

Engelbrecht C, Engelbrecht F. 2016. Shifts in Köppen-Geiger climate zones over southern Africa in relation to key global temperature goals. Theoretical and Applied Climatology **123,** 247-61.

Garnier E, Laurent G. 1994. Leaf anatomy, specific mass and water content in congeneric annual and perennial grass species. New Phytologist **128,** 725-736.

Geyer U, Schönherr J. 1990. The effect of the environment on the permeability and composition of *Citrus* leaf cuticles. I. Water permeability of isolated cuticular membranes. Planta **180,** 147-153.

Ghouil H, Montpied P, Epron D, Ksontini M, Hanchi B, Dreyer E. 2003. Thermal optima of photosynthetic functions and thermostability of photochemistry in cork oak seedlings. Tree Physiology **23,** 1031-1039.

Gil-Pelegrín E, Saz MA, Cuadrat JM, Pegueiro-Pina JJ, Sancho-Knapik D. 2017. Oaks under Mediterranean-type climates: functional responses to summer aridity. In: Gil-Pelegrín E, Peguero-Pina J, Sancho-Knapik D (eds) Oaks Physiological Ecology. Exploring the Functional Diversity of Genus *Quercus* L. Tree Physiology **7**, 137-192.

Guehl JM, Aussenac G. 1987. Photosynthesis decrease and stomatal control of gas exchange in *Abies alba* Mill. in response to vapour pressure difference. Plant Physiology **83**, 316-322.

Havaux M. 1993. Rapid photosynthetic adaptation to heat stress triggered in potato leaves by moderately elevated temperatures. Plant, Cell and Environment **46**, 461-467.

Himrane H, Camarero JJ, Gil-Pelegrín E. 2004. Morphological and eco-physiological variation of the hybrid oak *Quercus subpyrenaica* (*Q. faginea* × *Q. pubescens*). Trees **18**, 566-575.

Holloway PJ. 1982. The chemical constitution of plant cutins. In: Cutler DF, Alvin KL, Price CE (eds) The plant cuticle. Academic Press, London, UK, pp 45-85.

Huang H. 2017. Comparative investigation of the chemical composition and the water permeability of fruit and leaf cuticles. PhD book. University of Wüzburg, Germany.

IPCC. 2014. Climate change: mitigation of climate change. Contribution of Working Group III to the Fifth Assessment Report of the Intergovernmental Panel on Climate Change (ed. by Edenhofer O, Pichs-Madruga R, Sokona Y, Farahani E, Kadner S, Seyboth K, Adler A, Baum I, Brunner S, Eickemeier P, Kriemann B, Savolainen J, Schlömer S, von Stechow C, Zwickel T, and Minx JC). Cambridge University Press, Cambridge, UK and New York, NY, USA), 1-1435.

Iyengar BTR, Schlenk H. 1969. Melting points of synthetic wax esters. Lipids **4**, 28-30.

Jeffree CE. 2006. The fine structure of the plant cuticle. In: Riederer M, Müller C (eds) Biology of the plant cuticle: annual plant reviews, Vol. 23. Oxford: Blackwell Publishing, pp 11-125

Jetter R, Kunst L, Samuels AL. 2006. Composition of plant cuticular waxes. In: Riederer M, Müller C (eds) Biology of the plant cuticle: annual plant reviews, Vol. 23. Oxford: Blackwell Publishing, pp 145-181.

Jetter R, Riederer M. 2016. Localization of the transpiration barrier in the epi- and intracuticular waxes of eight plant species: water transport resistances are associated with fatty acyl rather than alicyclic components. Plant Physiology **170**, 921-934.

Karbulková J, Schreiber L, Macek P, Santrucek J. 2008. Differences between water permeability of astomatous and stomatous cuticular membranes: effects of air humidity in two species of contrasting droughtresistance strategy. Journal of Experimental Botany **59**, 3987-3995.

Kim KS, Park SH, Jenks MA. 2007. Changes in leaf cuticular waxes of sesame (*Sesamum indicum* L.) plants exposed to water deficit. Journal of Plant Physiology **164**, 1134-1143.

Knight CA, Ackerly DD. 2003. Evolution and plasticity of photosynthetic thermal tolerance, specific leaf area and leaf size: congeneric species from desert and coastal environments. New Phytologist **160**, 337-347.

Koch K, Hartmann KD, Schreiber L, Barthlott W, Neinhuis C. 2006. Influences of air humidity during the cultivation of plants on wax chemical composition, morphology and leaf surface wettability. Environmental and Experimental Botany 56, 1-9.

Körner C. 1995. Leaf diffusive conductances in the major vegetation types of the globe. In: Schulze ED, Caldwell MM (eds) Ecophysiology of photosynthesis. Springer, Berlin, pp 463-490.

Kosma DK, Bourdenx B, Bernard A, Parsons EP, Lü S, Joubès J, Jenks MA. 2009. The impact of water deficiency on leaf cuticle lipids of *Arabidopsis*. Plant Physiology **151**, 1918-1929.

Kosma DK, Jenks MA. 2007. Eco-physiological and molecular-genetic determinants of plant cuticle function in drought and salt stress tolerance. In: Jenks MA, Hasegawa PM, Jain SM (eds) Advances in molecular breeding toward drought and salt tolerant crops. Springer, Dordrecht, Netherlands, pp 91-210.

Ladjal M, Epron D, Ducrey M. 2000. Effects of drought preconditioning on thermotolerance of photosystem II and susceptibility of photosynthesis to heat stress in cedar seedlings. Tree Physiology **20**,1235–1241.

Lange OL. 1959. Untersuchungen über Wärmehaushalt und Hitzeresistenz mauretanischer Wüsten-und Savannenpflanzen. Flora **147**, 595-651.

Le Provost G, Domergue F, Lalanne C, Campos PR, Grosbois An, Bert D, Meredieu C, Danjon F, Plomion C, Gion J. 2013. Soil water stress affects both cuticular wax content and cuticle-related gene expression in young saplings of maritime pine (*Pinus pinaster* Ait). BMC Plant Biology **13**, 95.

Leide J, Hildebrandt U, Reussing K, Riederer M, Vogg G. 2007. The developmental pattern of tomato fruit wax accumulation and its impact on cuticular transpiration barrier properties: effects of a deficiency in a beta-ketoacyl-coenzyme A synthase (LeCER6). Plant Physiology **144,** 1667-1679.

Leide J, Hildebrandt U, Vogg G, Riederer M. 2011. The positional sterile (ps) mutation affects cuticular transpiration and wax biosynthesis of tomato fruits. Journal of Plant Physiology **168,** 871-877.

Leliaert F, Verbruggen H, Zechman FW. 2011. Into the deep: new discoveries at the base of the green plant phylogeny. Bioessays **33**, 683-692.

Lewis LA, McCourt RM. 2004. Green algae and the origin of land plants. American Journal of Botany **91**, 1535-1556.

Lihavainen J, Ahonen V, Keski-Saari S, Sober A, Oksanen E, Keinänen M. 2017. Low vapor pressure deficit reduces glandular trichome density and modifies the chemical composition of cuticular waxes in silver birch leaves. Tree Physiology **37**, 1166-1181.

Lo Gullo MA, Salleo S. 1988. Different strategies of drought resistance in three Mediterranean sclerophyllous trees growing in the same environmental conditions. New Phytologist **108,** 267-276.

Müller HM, Schäfer N, Bauer H, Geiger D, Lautner S, Fromm J, Riederer M, Bueno A, ... and Hedrich R. 2017. The desert plant *Phoenix dactylifera* closes stomata via nitrate-regulated SLAC1 anion channel. New Phytologist **216,** 150-162.

Niklas KJ. **1997**. The evolutionary biology of plants. Chicago: University of Chicago Press.

Niklas KJ. 2016. Plant evolution: an introduction to the history of life. Chicago: University of Chicago Press.

Nobel PS. 2009. Physicochemical and environmental plant physiology, 4th edn. Academic Press, Oxford.

Oliveira AF, Salatino A. 2000. Major constituents of the foliar epicuticular waxes of species from the Caatinga and Cerrado. Zeitschrift für Naturforschung C **55**, 688-692.

Panahi P, Jamzad Z, Pourmajidian MR, Fallah A, Pourhashemi M. 2012. Foliar epidermis morphology in *Quercus* (subgenus *Quercus*, section *Quercus*) in Iran. Acta Botanica Croatica **71**, 95-113.

Peel MC, Finlayson BL, McMahon TA. 2007. Updated world map of the Köppen-Geiger climate classification. Hydrology and earth system sciences discussions **4**, 439-473.

Peguero-Pina JJ, Morales F, Flexas J, Gil-Pelegrín E, Moya I. 2008. Photochemistry, remotely sensed physiological reflectance index and de-epoxidation state of the xanthophyll cycle in *Quercus coccifera* under intense drought. Oecologia **156**, 1-11.

Peguero-Pina JJ, Sisó S, Fernández-Marín B, Flexas J, Galmés J, García-Plazaola JI, Niinemets U, Sancho-Knapik D, Gil-Pelegrín E. 2016. Leaf functional plasticity decreases the water consumption without further consequences for carbon uptake in *Quercus coccifera* L. under Mediterranean conditions. Tree Physiology **36**, 356-367.

Pensec F, Pączkowski C, Grabarczyk M, Woźniak A, Bénard-Gellon M, Christophe Bertsch C, Julie Chong J, Szakiel A. 2014. Changes in the triterpenoid content of cuticular waxes during fruit ripening of eight grape (*Vitis vinifera*) cultivars grown in the upper Rhine Valley. Journal of Agricultural and Food Chemistry **62**, 7998-8007.

Perkins SE, Alexander SV, Naim JR. 2012. Increasing frequency, intensity and duration of observed global heatwaves and warm spells. Geophysical Research Letters **39**, L20714.

Pollard M, Beisson F, Li Y, Ohlrogge JB. 2008. Building lipid barriers: biosynthesis of cutin and suberin. Trends in Plant Science **13**, 236-246.

Purves WK, Sadava D, Orians GH, Heller HC. 2004. Life: the science of biology. 7th ed. Sunderland: Sinauer Associates, 772.

Raven PH, Evert RF, Eichorn SE. 2005. Biology of Plants. WH Freeman, New York, USA.

Reich PB, Walters MB, Ellsworth DS. 1997. From tropics to tundra: global convergence in plant functioning. Proceedings of the National Academy of Sciences **94,** 13730-13734.

Reichardt PB, Bryant JP, Clausen TP, Wieland GD. 1984. Defense of winter dormant Alaska paper birch against snowshoe hares. Oecologia **65**, 58-69.

Retief JJ, le Roux JH. 1983. Crystallographic investigation of a paraffinic Fischer-Tropsch wax in relation to a theory of wax structure and behavior. South African Journal of Science **79**, 234-239.

Reynhardt EC, Riederer M. 1994. Structure and molecular dynamics of plant waxes. European Biophysics Journal **23,** 59-70.

Riederer M, Schneider G. 1990. The effect of the environment on the permeability and composition of *Citrus* leaf cuticles. Planta 180, 154-165.

Riederer M, Schönherr J. 1988. Development of plant cuticles: fine structure and cutin composition of *Clivia miniata* Reg. leaves. Planta **174**, 127-138.

Riederer M, Schreiber L. 1995. Waxes - the transport barriers of plant cuticles. In: Hamilton RJ (ed) Waxes: chemistry, molecular biology and functions. Dundee: The Oily Press, pp 131-156.

Riederer M, Schreiber L. 2001. Protecting against water loss: analysis of the barrier properties of plant cuticles. Journal of Experimental Botany **52**, 2023-2032.

Riederer M. 1991. Cuticle as barrier between terrestrial plants and the atmosphere-significance of growth-structure for cuticular permeability. Naturwissenschaften **78**, 201-208.

Riederer M. 2006. Thermodynamics of the water permeability of plant cuticles: characterization of the polar pathway. Journal of Experimental Botany **57**, 2937-2942.

Roth-Nebelsick A, Fernández V, Peguero-Pina JJ, Sancho-Knapik D, Gil-Pelegrín E. 2013. Stomatal encryption by epicuticular waxes as a plastic trait modifying gas exchange in a Mediterranean evergreen species (*Quercus coccifera* L.). Plant, Cell and Environment **36**, 579-589.

Rubio de Casas R, Vargas P, Pérez-Corona E, Manrique E, Quintana JR, García-Verdugo C, Balaguer L. 2007. Field patterns of leaf plasticity in adults of the long-lived evergreen *Quercus coccifera*. Annals of Botany **100**, 325-334.

Scareli-Santos C, Sánchez-Mondragón ML, González-Rodríguez A, Oyama K. 2013. Foliar micromorphology of Mexican oaks (*Quercus*: Fagaceae). Acta Botanica Mexicana **104**, 31-52.

Schmidt HW, Schönherr J. 1982. Development of plant cuticles: occurrence and role on non-ester bonds in cutin of *Clivia miniata* Reg. leaves. Planta **156**, 380-384.

Schönherr J, Lendzian K. 1981. A simple and inexpensive method of measuring water permeability of isolated plant cuticular membranes. Zeitschrift für Pflanzenphysiologie **102**, 321-327.

Schönherr J, Riederer M. 1986. Plant cuticle sorb lipophilic compounds during enzymatic isolation. Plant, Cell and Environment **9**, 459-466.

Schönherr J. 1976. Water permeability of isolated cuticular membranes: the effect of cuticular waxes on diffusion of water. Planta **131,** 159-164.

Schreiber L, Riederer M. 1996. Ecophysiology of cuticular transpiration: comparative investigation of cuticular water permeability of plant species from different habitats. Oecologia **107,** 426-432.

Schreiber L. 2001. Effect of temperature on cuticular transpiration of isolated cuticular membranes and leaf discs. Journal of Experimental Botany **52,** 1893-1900.

Schreiber, L. 2002. Co-permeability of ^{3}H-labelled water and ^{14}C-labelled organic acids across isolated *Prunus laurocerasus* cuticles: effect of temperature on cuticular paths of diffusion. Plant Cell and Environment **25,** 1087-1094.

Schulze ED, Beck E, Müller-Hohenstein K. 2005. Plant Ecology. Springer, Heidelberg-Berlin, Germany.

Schuster A-C, Burghardt M, Alfarhan A, Bueno A, Hedrich R, Leide J, Thomas J, Riederer M. 2016. Effectiveness of cuticular transpiration barriers in a desert plant at controlling water loss at high temperatures. AoB Plants 8:plw027 doi:10.1093/aobpla/plw027.

Schuster A-C, Burghardt M, Riederer M. 2017. The ecophysiology of leaf cuticular transpiration: are cuticular water permeabilities adapted to ecological conditions? Journal of Experimental Botany **68,** 5271-5279.

Schuster A-C. 2016. Chemical and functional analyses of the plant cuticle as leaf transpiration barrier. PhD book. University of Würzburg, Germany.

Si Y, Zhang C, Meng S, Dane F. 2009. Gene expression changes in response to drought stress in *Citrullus colocynthis*. Plant Cell Reports **28,** 997-1009.

Slavík B. 1974. Methods of studying plant water relations. Springer, Berlin, Germany.

Small DM. 1986. The physical chemistry of lipids. From alkanes to phospholipids. New York, London: Plenum Press.

Smith SD, Monson RK, Anderson JE. 1997. Physiological ecology of North American desert plants. Springer, Berlin-Heidelberg, Germany.

Szakiel A, Niżyński B, Pączkowski C. 2013. Triterpenoid profile of flower and leaf cuticular waxes of heather *Calluna vulgaris*. Natural Product Research **27,** 1404-1407.

Teskey R, Wertin T, Bauweraerts I, Ameye M, McGuire MA, Steppe K. 2015. Responses of tree species to heat waves and extreme heat events. Plant, Cell and Environment **38**, 1699-1712.

Trenberth KE, Dai A, Van Der Schrier G, Jones PD, Barichivich J, Briffa KR, Sheffield J. 2014. Global warming and changes in drought. Nature Climate Change **4**, 17-22.

Tsubaki S, Sugimura K, Teramoto Y, Yonemori K, Azuma J. 2013. Cuticular membrane of *Fuyu persimmon* fruit is strengthened by triterpenoid nano-fillers. PLoS One **8**:e75275.

Vilagrosa A, Bellot J, Vallejo VR, Gil-Pelegrín E. 2003. Cavitation, stomatal conductance, and leaf dieback in seedlings of two co-occurring Mediterranean shrubs during an intense drought. Journal of Experimental Botany **54**, 2015-2024.

Vogg G, Fischer S, Leide J, Emmanuel E, Jetter R, Levy AA, Riederer M. 2004. Tomato fruit cuticular waxes and their effects on transpiration barrier properties: functional characterization of a mutant deficient in a very-long-chain fatty acid β-ketoacyl-CoA synthase. Journal of Experimental Botany **55**, 1401-1410.

Volder A, Tjoelker MG, Briske DD. 2010. Contrasting physiological responsiveness of establishing trees and a C_4 grass to rainfall events, intensified summer drought, and warming in oak savanna. Global Change Biology **16**, 3349-62.

Ward D. 2016. The biology of deserts. Oxford University Press, Oxford, United Kingdom.

Waters ER. 2003. Molecular adaptation and the origin of land plants. Molecular Phylogenetics and Evolution **29**, 456-463.

Weis E, Berry JA. 1988. Plants and high temperature stress. In: Long SP, Woodward FI, eds. Plants and temperature. Cambridge, UK: The Company of Biologists Limited, 329-346.

Wright IJ, Reich PB, Westoby M *et al.* 2004. The worldwide leaf economics spectrum. Nature **428,** 821-827.

Yeats TH, Rose JKC. 2013. The formation and function of plant cuticles. Plant Physiology **163**, 5-20.

Zhu L, Bloomfield KJ, Hocart CH, Egerton JJ, O'Sullivan OS, Penillard A, Weerasinghe LK, Atkin OK. 2018. Plasticity of photosynthetic heat tolerance in plants adapted to thermally contrasting biomes. Plant, Cell and Environment **41**, 1251-1262.

Zobayed SMA, Afreen F, Kubota C, Kozai T. 2000. Water control and survival of Ipomoea batatas grown photoautotrophically under forced ventilation and photomixotrophically under natural ventilation. Annals of Botany **86**, 603-610.

www.ingramcontent.com/pod-product-compliance
Lightning Source LLC
LaVergne TN
LVHW041702190726
843493LV00007B/1915